# THE MANY LIVES OF CARBON

REAKTION BOOKS

# THE MANY LIVES OF CARBON

THE MANY LIVES OF CARBON

DAG OLAV HESSEN

Published by
REAKTION BOOKS LTD
Unit 32, Waterside
44–48 Wharf Road
London N1 7UX, UK
www.reaktionbooks.co.uk

First published by Reaktion Books in 2017
English translation by Kerri Pierce

Originally published by Cappelen Damm AS, Oslo

This translation has been published with the financial support of NORLA



Printed and bound in Great Britain
by TJ International, Padstow, Cornwall

A catalogue record for this book is available from the British Library

ISBN 978 1 78023 851 7

# CONTENTS

PART II

# THE C IN CYCLE

95

PART III

# THE FOOTPRINT

# PART I

# CARBON, CARBON EVERYWHERE

Heavy drops fall from the surrounding trees and rain splashes out onto the water. It also trickles down a tarpaulin spread between the trees. Beneath the tarpaulin three people are bent over a microscope, pipettes, small plastic boxes, filtering tools and other things belonging to a lab. The fog around us is dense, it is midnight, dark, and we work on our knees in the light of headlamps. The samples are placed in small plastic bottles, which are then filled with a rather viscous, transparent liquid from a metal container with a dispenser. The caps are screwed on tightly, the samples are marked and placed in a rack. Every four hours two of us set out in a small, flimsy boat to retrieve more samples from six large bags attached to a framework. Each bag holds a couple of thousand litres of water. Then we row back, crawl beneath the tarpaulin and persevere . . .

Around dawn the rain lessens. With a foggy mind in foggy surroundings I sit at the water's edge and watch rays of life penetrate the morning mist, which is now becoming lighter over the water. Two dragonflies patrol the water lilies along the shore. The forest around the water is dense and green. Suddenly, I have a sense that nature is breathing all around me. Everything seethes with life, and it seethes with *carbon*. Except for the water's smooth surface and the sky itself, everything around me is shades of green: billions of chloroplasts. Carbon, in the form of carbon dioxide ($CO_2$), is breathed in and out through the endless array of stomata in leaves and needles before it is transported to chloroplasts. There the world's most important chemical reaction occurs, where the gas is almost miraculously transformed into organic carbon in the form of carbohydrates like sugar, starch and cellulose. As a bonus, a formidable bonus, that same reaction produces the oxygen upon which the other half of planetary life depends. And this plant-bound carbon makes its way into ecosystems, on which we all depend.

Carbon is chemistry's potato, a basic ingredient in nearly all the body's molecules and components. Indeed, truth be told, carbon occurs in almost everything – in the most amazing varieties. For all that it is common, carbon is indispensable. I'm conducting my own experiment in large plastic containers out on the lake, and every piece of equipment I can think of consists of carbon polymers: pipettes, containers, the racks in which they stand, the cap over my head, even my clothes themselves. Is there anything here that is *not* carbon? And then my own carbon, which I have only borrowed, which is making a simple detour as it follows me through life, has circulated in plants, bacteria, animals and minerals for billions of years, and shall continue to do so billions of years after I am gone. As participants in the carbon cycle we all have a share in eternity.

There at the water's edge, thoughts of the carbon cycle, and my own, albeit fleeting and humble, participation in it gives me, from an infinite perspective, a sense of belonging. Carl Linnaeus (Linné) believed that we humans were placed here on earth as interpreters of divine Creation.[1] Unabashedly, he regarded himself as among the foremost of these. We three next to a lake in the woods are on a more modest errand, but an errand that will nonetheless turn out to be more significant than we thought: we are here to understand more about carbon's eternal ring dance.

The six gigantic bags out on the water have been treated with minute quantities of a special carbon, namely radioactive carbon, the isotope 14C, which in contrast to the normal form, 12C, contains six protons and eight neutrons (6 + 8 = 14). The added $14CO_2$ makes it possible for us to trace how carbon is taken up by algae, water's single-celled plants, into bacteria, into zooplankton, and out into the system again as $14CO_2$. The radioactive carbon in our samples, contained in the small plastic bottles, sheds electrons that emit a tiny light signal every time they are caught by the viscous liquid. These glints can be captured when we later place the samples into a scintillation counter. This will become my doctoral thesis and it will enable us to understand a *little* more about carbon. That night beneath the tarpaulin, together with samples taken on countless other days and nights, not to mention an unknown number of analyses, statistical treatments and written work, will become a central part of the thesis regarding carbon's inscrutable pathways. Radioactive carbon was the key to unlocking photosynthesis itself, and here there is a direct connection from nuclear physical to biological insights.

The sun, the motor driving it all, rises. It grows hot and the mist disappears altogether. Before the samples are placed in the scintillation counter, I will permit myself a few hours of sleep. Watching results tick out of measuring instruments is a special joy to a researcher, one that can only seldom be shared with the rest of the world. Sometimes it takes only a day to find out whether the points on the curve correspond to expectations. Other times it can take years and requires a rare kind of patience that today's researchers seldom possess or can allow themselves to have. Still, Charles David Keeling had what it took. In 1957 he insisted on the ridiculous notion of placing a measuring instrument on top of Hawaii's Mauna Loa.[2] The resulting figures led to a greater awareness surrounding carbon than my own, even if, in all humility, they belong to the same story – the story of carbon. Thanks to his unusual combination of precision and patience,

Keeling, much to his dismay, was able to confirm that $CO_2$ concentrations were rising year after year. That could only indicate a fundamental imbalance in the carbon cycle: more $CO_2$ was being released than was being absorbed. Life-giving $CO_2$, therefore, had a darker side. Eventually, it was shown that the concentration of methane in the atmosphere was undergoing the same disquieting increase. Carbon, life's element, has become our greatest threat.

Most life processes taking place inside and around us involve carbon in one form or another. Carbon also has many lives of its own. In pure form, it can appear as pencil 'lead' (graphite) or diamonds. Simultaneously, it forms the springboard for most of the synthetics upon which modern life is based, though the boundary between natural and synthetic carbon is not as sharp as one would think. Yet the heart of this story is carbon's life cycle or cycles, the magnificent balance between photosynthesis and cellular respiration, between what it is to build and what it is to burn. When it comes to relations with other atoms, carbon is a first star, and one of the most intense and pivotal relationships into which C enters is that with two Os. $CO_2$ is the gas of life and death, and in order to understand the climate's future and past development we must understand the carbon cycle and how we affect it. The planet's future depends upon our interaction with carbon.

We will duly return to the carbon cycle(s), to Keeling himself and his famous curve, with all that it implies, but let us now take a closer look at the main protagonist in this story, carbon, and its numerous, if not to say countless, variations, as well as the alliances it forms.

I have a close relationship to my subject. We all do, if only unconsciously. My life as a researcher has largely revolved around carbon, in everything from the contents of the nucleus to the great cycle contained in ecosystems. I know my object from the worlds of both chemistry and biology, and I have tried my best to gain insight into carbon's many partners, life phases and countless roles. Most of all, however, I want to understand carbon's past and future role in our climate.

## CARBON IN EVERYTHING

Carbon is stardust, formed in stars by lighter elemental fusion. It is everywhere in the universe, but only thinly distributed. Carbon accounts for less than five parts per thousand of the universe's total elemental mass, yet is nonetheless the fourth most common element. In the earth's atmosphere, its weight is just under 400 parts per million (ppm), while on the earth itself it amounts to even less, only 200 ppm.[3]

Life, though, shows an affinity for carbon. In a human being, the top ten elements, based on their weight – which will actually be about the same in a mouse or an elephant or any other animal – is as follows: oxygen (65 per cent), carbon (18 per cent), hydrogen (10 per cent), nitrogen (3 per cent), calcium (1.5 per cent), phosphorus (1 per cent), potassium (0.35 per cent), sulphur (0.25 per cent), sodium (0.15 per cent) and magnesium (0.05 per cent).[4] If I discount water, which makes up about 57 per cent of me, and accounts for much of the oxygen and hydrogen, then I am over 40 per cent carbon. Carbon is, so to speak, involved in all of my body's organization and upkeep, and in many ways the body's budget can be calculated according to the following equation: ingested carbon minus undigested carbon minus $CO_2$ from cellular respiration = net weight gain or weight loss.

Carbon is the central element in the main groups of molecules (DNA, proteins, fat, sugars) that compose all living organisms. Animals are about 40 per cent carbon, if we discount the body's water content, depending upon the ratio between fat, carbohydrates and proteins. Plants hold somewhat more carbon, especially trees with their significant cellulose and lignin content. Plants have also ensured that carbon has enjoyed its own geological era, the Carboniferous Period. This occurred from 360 to 300 million years before our present era, and is named for the vast forests that thrived in a warm and moist climate. The coal left by these enormous forests, accumulated through 60 million years, is the origin of carbon – both as a geological epoch and otherwise. The much later Cretaceous Period, derived from Latin *creta* for 'chalk', which lasted from 145 to 65 million years before our present era, is also characterized by its carbon deposits, although in the form of calcium carbonate, $CaCO_3$.[5] Here carbon has bonded to one calcium atom and three oxygen atoms, an alliance enabled by billions upon billions of microscopic, lime-forming algae. We shall hear more about this later.

A few hundred million years, of course, are nothing to the carbon atoms found in coal and chalk. When the universe was formed 14 billion years ago, the first elements born were the lightest three: hydrogen, helium and lithium. Somewhat simplified, we can say that these proved the building blocks for heavier elements like carbon, which also ranks fourth as the universe's most common element (by weight), after hydrogen, helium and oxygen.

When it comes to the periodic table, C enjoys boundless popularity among countless potential partners, and it has brief or long-lasting affairs with a significant number of other atoms – one or many at a time. The carbon atom's popularity is due to its unusual propensity for forming chemical bonds. Almost everything is attracted to C, mostly because of its six electrons and six protons. There are also six neutrons, and carbon's six neutrons and six protons are what gives it an atomic weight of 12. The weight of its electrons is insignificant in this context, but becomes more important in other contexts. Since two of these electrons, moreover, are hidden away in C's inner electron shell, it is the four electrons remaining in the outer shell that seek contact with the outside world, and they do so with resounding success.

In nature carbon occurs in three main forms: graphite, diamond and fullerene – a football-like, hollow molecule of carbon. In each of these forms, carbon atoms are busy interacting and have formed a wide variety of different bonds with each other. At the same time, C is not picky when it comes to forming friendships with other atoms. Indeed, one of C's favourite relationships is a three-way bond with two O atoms. It is this molecule, $CO_2$, that transforms C into life's atom, since almost all life is based upon it. However, it is also problematic that C exhibits such a fidelity to this bond and has a tendency to remain there. Combustion more than anything cements these atoms together, whereas the intricate mechanism involved in photosynthesis is able to separate the trio. Today the balance between these processes is so substantially askew that $CO_2$ is accumulating in the atmosphere, and that is the reason why C's affinity for O is a relationship that affects the entire world.

Carbon is significantly older than the earth (which has 'only' been around for about 4 billion years), but it is younger than the

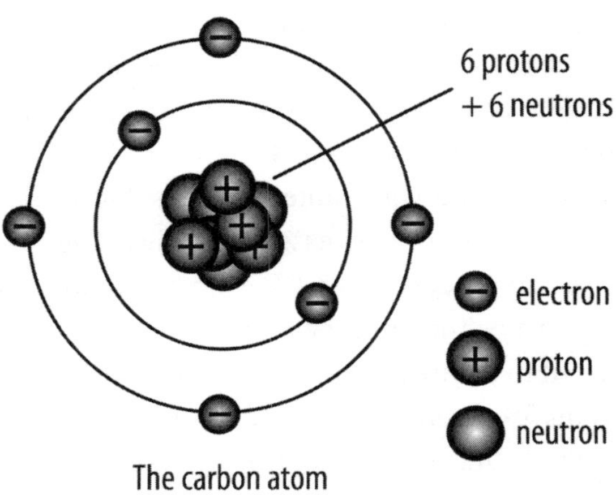


The attractiveness of the carbon atom is largely due to its six electrons, while the atomic weight of 12 is due to the six neutrons and six protons.

universe. Nonetheless, it was circulating in and out of stars and chemical relationships for billions of years before it materialized in the first bacteria, in trees, in dinosaurs, in you and in me. Even spending a few hundred million years out of circulation, in the form of coal, oil or gas, is not terribly significant compared to a carbon atom's eternal life cycle. Still, how long have *we* been aware of carbon?

When it comes to scientific discoveries, it is not unusual to refer back to 'the ancient Greeks' as the origin of all knowledge. When it comes to carbon, however, we can effortlessly travel even farther back. Of course, it would take a fair amount of human history before we were able to understand combustion and explain physics and chemistry, but ever since we marvelled at fire's magic and its potentially destructive power, back in some dim prehistoric age when we slunk around the savannah with our wooden clubs, the trio of $CO_2$ was recognized and to a certain extent utilized. Even the word *carbon* is Latin for 'burnt wood' (*carbo*). When it came to characterizing nature early on, one clear distinction could be made: there were things like wood, which burned, and things like rocks and sand, which did not. Fire's logical opposite was water, which proved fire's enemy. Little did anyone know that what nourished fire and what extinguished it both largely consisted of oxygen, or that the end product produced by burning was largely $CO_2$.

The fact that the very same reaction that enabled fire to warm us from without was essentially the same process our bodies use to supply energy and life from within would have been astounding, if not incomprehensible. In both cases, the solar energy stored in plants is released as $CO_2$. Fire itself is ancient, and the campfire is probably as old as our ancestors' ability to walk upright. Sitting at the campfire, ideally with a view, ideally with something at one's back, and most ideally next to water, is a ritual almost imprinted in our genes. Almost everywhere matching that description a stone, ash-filled ring may be found, demonstrating one of those unbroken cultural lines extending from the origin of our species to today. We cannot say for sure whether our ancestors, as they sat next to a fire, were much puzzled by how a cold piece of wood suddenly could be transformed to intense heat and then to ash. But we can presume that someone sitting around that circle probably was. Still, even though fire proved a central component of our prehistory, both as friend and foe, its actual essence remained a mystery, as did its ability to create heat from cold.

Fire and combustion are crucial not just for joining carbon to oxygen, but also to the whole carbon cycle, not to mention to the transitions between various forms of carbon. Fire has also proven central to our own combustion and evolution.[6] The human brain's development has been remarkably quick, in respect to both size and function. Our nearest living relative, the chimpanzee, has a brain volume between 300 and 500 $cm^3$, and our larger, but more distant relative, the gorilla, varies between 400 and 700 $cm^3$. Our early ancestor *Homo habilis*, who lived around 1.7 million years ago, had a brain volume akin to a gorilla's, but from there development took off. Only 700,000 years later did *Homo erectus* approach 1,000 $cm^3$, while both *Homo neanderthaliensis* and *Homo sapiens* were nearing a maximum brain volume of around 1,500 $cm^3$. Since then the brain's development has levelled out, even decreased a little, and Neanderthals actually had a greater brain volume than us. Still, size is not everything that matters; we also have a remarkably high density of neurons and good insulation around the synapses, where we find modern brain sections like the frontal lobes, which contain much of our sense of self, as well as the ability to make abstract reflections and ethical judgements.

Yet this formidable brain capacity also exacts a price in the form of energy. More than 20 per cent of our energy expenditure is spent on the brain, an organ that, despite everything, amounts to only 2 per cent of our body weight. In well-nourished parts of the world today this energy drain does not present a problem for us, but it exacts a hefty cost in other parts of the world. Historically speaking, however, it has been a high price that was obviously worth paying.

Yet where do the campfire and carbon enter the picture? The answer, Richard Wrangham claims, is found in the title of his book, *Catching Fire: How Cooking Made Us Human*.[7] We have probably been using fire for around 100,000 years, and even before then our ancestors could observe after a savannah fire it was simple to gather dead game that was both thoroughly cooked and much easier to eat. That presumably gave them the idea to use fire more strategically to hunt or trap wild quarry inside, and since all animals fear fire, a burning branch could keep an otherwise terrifying lion at bay. Fire changed from foe to friend. Roasting, and eventually boiling, was also excellent at killing the parasites and bacteria found on and in food. Fire not only changed the chemistry of food, but also its – and our own – biology. Roasted or boiled food is easier to chew, something that makes extensive chewing muscles and a massive jaw redundant, and the food is also easier to digest. It is difficult to get enough energy with raw fare, and whereas other primates must spend their days trying to consume at least 20 to 30 per cent of their body weight just to cover their energy requirements, we manage just fine with 5 per cent. Foodstuffs such as wheat, rice and potatoes are, practically speaking, indigestible in a raw state. But with the addition of fire, they became central components of our diet. As a result, fire was probably a significant step on the road to the modern human and our dominion over nature – parts of it anyway. Not only was it now possible to amass enough energy to maintain our large brain and manage with a simpler digestive system, but we were freed to cultivate social relationships (another contributor to a larger brain), solve practical problems and spend energy puzzling about how things burned.

Maybe 50,000 years ago a youth was already sitting by the remnants of fire after everyone else in the group, stomachs full of roast boar, had gone to sleep, speculating on where the wood had gone. All that remained, after all, was some blackened debris and a little ash. Perhaps he wondered why one's fingers became black after touching something fire had left. Perhaps he experimentally stroked his arm and made some streaks on a stone. Perhaps it was after just such a philosophical moment at the campfire that an early European took a piece of charcoal from the fire pit and sketched the first stylized mammoth on a cave wall in utter gratitude for fire and life.

CHARCOAL ITSELF is old news. It is a form of carbon that has been familiar for as long as intelligent humankind (that is, *Homo sapiens*) has been around. It has also been employed in the service of art since the first scratches on a cave wall were made 40,000 years ago. Around this time an assorted selection of animals, rather impressively true to life, or at least their outlines (the artist's signature?), were created using charcoal alone or with various pigmentations. Indeed, the fact that identical art from the same epoch is found on cave walls both in Europe and Indonesia might also indicate that art was already established before that time, and was brought with immigration from Africa. Still, how can we know that these paintings originated 40,000 years ago? It is here that carbon's radioactive component comes into play, since it can be used for carbon dating.

Carbon dating is based on the fact that a small amount of the radioactive carbon isotope 14C is found in all living organisms. As long as the organism is alive, the ratio of radioactive 14C to 'normal' C (12C) will remain constant. When the organism dies, however, the ratio of 14C will decrease with a half-life of 5,730 years. By measuring the amount of radioactivity in relation to fresh organic material, one can calculate, for example, how long ago the tree that produced the said charcoal grew: 40,000 years, however, is about the maximum time period for this method. Willard F. Libby, who described this particular method of carbon dating in 1949, was awarded a well-earned Nobel Prize for the discovery in 1960.[8] Archaeologists and others wanting to date ancient relics now had a firm foundation beneath their feet, as long as their finds held carbon, which they most certainly did.

Cave art in Indonesia has long been well known, but it was only in 2014 that it could be proven that the artist had not accomplished his work 10,000 years ago, but rather 40,000 years ago. In this case, there was not enough organic carbon left for 14C-dating, but carbon nonetheless provided aid: in the limestone cave, nodules of calcite or calcium carbonate – a bond between carbon, calcium and oxygen ($CaCO_3$) – had formed on top of parts of the paintings. Calcite traps uranium, which has a recognized half-life of several million years. With the help of a special saw outfitted with diamond teeth, the nodules could be parted horizontally with surgical precision, allowing the changes in uranium isotopes on the oldest deposits topping the cave art to be measured and thereby to calculate the paintings' age in terms of the changes in relationship between two uranium isotopes. These were found to be even older than the European cave art, which generally dates between 20,000 and 32,000 years ago.[9] The Chauvet Cave in southern France, which has been calculated up to 35,000 years with the help of 14C, was the earliest known. Now there are also claims of 40,000-year-old cave art in the form of handprints in a cave in Malaga, Spain. Could it have been Neanderthals? The jury is still out, in terms of both the art's age and origin – if indeed it is art. Nonetheless, carbon's role in art is an ancient one, and it is carbon itself that allows us to determine exactly *how* ancient that is.

## FIRE EXPLAINED

The fascination with fire, along with visible remnants such as charcoal, ash or soot, was understandable and had a practical side. Fire's invisible end product in the form of $CO_2$ was of more academic interest. Indeed, it was a pure mystery how a heavy wooden log could vanish into thin air, leaving only a thin ash layer behind that a breath of wind could lift.

Here we can also pass over the ancient Greeks, and make the leap from 40,000 years ago to Europe on the cusp of the Enlightenment. The person who first realized that fire, carbon and oxygen shared a common fate, and who described $CO_2$ as resulting from burnt charcoal, was Jan Baptiste van Helmont, born in Brussels in 1580.[10] He is also given the honour of introducing the word 'gas'. Helmont conducted experiments with plants, weighing the earth in which they grew and recording the difference in weight before and after the growing season. The earth weighed the same, but the plants had increased in weight. Where did that weight increase come from? Water, Helmont concluded. Obviously, the plants had received water and the water itself had disappeared. With the clarity of hindsight, one might lose a little respect for Helmont's conclusion, simply because $CO_2$ does not factor into the equation. And though everything is built on the basis of something else, it is unlikely that Helmont's discoveries were widely known. It is astonishing how often the wheel is reinvented, and back then scientific innovations could not be searched for and saved with a couple of keystrokes.

When it comes to the discovery of $CO_2$, among those fighting for a place on the podium was the Scotsman Joseph Black, born in 1728 and, from 1756, a professor of anatomy and chemistry.[11] Black astutely observed that adding acid to limestone produced a gas he called 'fixed air', and concluded that this must be the very gas we exhaled. One of the other pioneers who sought to enter the invisible world of gases was another Joseph, the Englishman

Joseph Priestley. Priestley was the same age as Black, and potentially a man of even more talents: theologian, philosopher, chemist and pedagogue. His ideas on utilitarianism inspired key thinkers such as Jeremy Bentham, John Stuart Mill and Herbert Spencer. Utilitarianism claims, in somewhat simplified term, that 'the greatest happiness of the greatest number' is the goal of all things: not happiness in a narrow or hedonistic sense, but rather the possibility of realizing one's abilities. Priestley, in any case, realized his own.

Whether or not Priestley knew of Black's discovery is unknown, but it was only after becoming inspired by conversations with Benjamin Franklin at the end of the eighteenth century that he seriously threw himself into science. Priestley confirmed that $CO_2$, which was produced, for example, through the process of beer fermentation, was lethal to fire and animals in high concentrations. He noted with interest that plants, in contrast, thrived on $CO_2$, and realized, unlike his predecessor Helmont, that $CO_2$ was the primary source of plant growth. With that he was on the trail of the great carbon cycle, the shifts between photosynthesis and cellular respiration. He also believed that the gas had nearly countless possible uses. In 1772 he wrote the article 'Directions for Impregnating Water with Fixed Air'.[12] In other words, the recipe for 'sparkling water', or mineral water, which he assumed would have significant uses, both medicinal and otherwise. Priestley could have laid the foundation for a significant

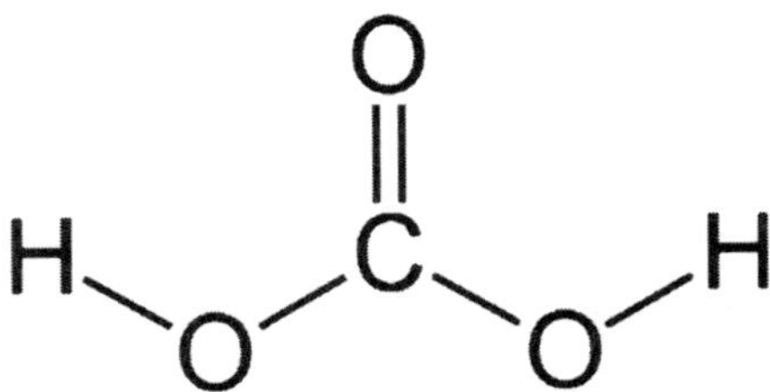


$H_2CO_3$ is perhaps the world's most common and important acid, although it is also one of the world's weakest.

industry here, but the person who took the idea further was a Swiss immigrant by the name of Johann Jacob Schweppe, who in 1792 established a soda water business in London. The company is still going strong, although today it is mostly known for soda and tonic water. Priestley obtained his $CO_2$ from fermentation (which is also combustion, the end product of yeast's cellular respiration, where sugar and starch are transformed to $CO_2$), although he did not connect $CO_2$ to fire except to note that in sufficient quantities $CO_2$ could quench fire.

Oxygen forms a weak acid when it first bonds with carbon to become $CO_2$ and then with water. So why does the compound $H_2CO_3$ become acidic? Because it is able to release the two hydrogen atoms as protons (H+), which is the very essence of an acid. There is an equilibrium between $CO_2$, water and carbonic acid, which is written as $CO_2 + H_2O \rightleftarrows H_2CO_3$. C, O and H can enter into an almost endless number of combinations, but carbonic acid is one of the most important, mainly for water. It is the reason fresh water in equilibrium with $CO_2$ is weakly acidic, and it is, as we will later see, an essential reaction in ocean acidification because greater amounts of $CO_2$ produce more $H_2CO_3$,which is acidic.

Many of the relatively loose threads relating to fire, air and carbon were woven together in Paris in the revolutionary year of 1789 in Antoine Lavoisier's pioneering chemistry textbook, *Traité élémentaire de chimie* (Elementary Treatise of Chemistry).[13] Lavoisier categorizes carbon here as a non-metallic element able to acidify and oxidize – that is, combust. But even though Lavoisier's book in many ways consolidated chemistry as a discipline, he stood on many other people's shoulders in his description of carbon. Just who should be acknowledged for what, however, is not entirely clear. Priestley not only discovered the existence and characteristics of $CO_2$, he is also attributed with discovering oxygen, even though it is an honour he probably shares with many others, among them Lavoisier and a Swede, Carl Wilhelm Scheele. In 1777, a full twelve years before Lavoisier's textbook, Scheele wrote his *Chemical Treatise on Air and Fire*, in which he essentially describes oxygen, carbon dioxide and nitrogen.[14] As far as oxygen is concerned, Scheele discovered the gas on 1 August 1772 by heating up mercury oxide. Today Scheele would have immediately published his findings, but things moved more slowly in the 1700s. In 1774 Scheele wrote a letter to Lavoisier, who was already established as the chemist par excellence, explaining his experiments and seeking his advice on further experimentation.

During this same period Priestley was conducting experiments that demonstrated how plants in a closed chamber, and beneath lighting, can 'purify' the air of $CO_2$. After the plants have done their job, fire can again burn and animals can again breathe in the chamber. Priestley also conducted experiments with heated mercury oxide in which he registered the release of oxygen. However, he was quicker to publish his results, which he did in 1775, and was therefore on the official record before Scheele. Like Scheele, Priestley consulted Lavoisier. He was not content with a mere letter, however, but instead travelled to Paris to discuss his experiments directly with the famous French chemist. Lavoisier listened with interest and immediately conducted his own experiments. He soon realized that oxygen was the essence of fire – flames, therefore, had no mythical or 'essential' substance as had been thought – and he calculated that the atmosphere had a 25 per cent oxygen content. That is impressively close to the actual number, which is 21 per cent. Should Lavoisier not have acknowledged Priestley's insights, not to mention Scheele's, who is never named despite his 1774 letter?

Lavoisier claimed that he had never received Scheele's letter, even though it was found among his wife's belongings many years later. Could it be that she never showed the letter to her husband? We will never know. We can give Lavoisier the benefit of the doubt, of course, since he was certainly a genius within chemistry. With impressive insight he compiled a list of nature's elementary building blocks, and he also realized, in contrast to Scheele and Priestley, what actually occurred in combustion: when oxygen bonded with carbon, yielding $CO_2$ as part of the equation, there would be a net weight increase, not a decrease, in terms of what was being burned. As mentioned, however, Lavoisier's 1789 chemistry textbook was published in the year the French Revolution erupted. Even if Lavoisier was a visionary when it came to chemistry, he was no political visionary and supported the royalist faction, although he did decline the position of finance minister to Louis XVI in 1791. Nonetheless, he was caught in the revolution's undertow and, like so many others, fell victim to the guillotine on 5 May 1794. The mathematician Joseph-Louis Lagrange attended the execution and was heard to remark laconically: 'It took them only an instant to cut off this head, and one hundred years might not suffice to reproduce its like.' Scheele himself did not enjoy a long life, but died in 1786 at only 43 years old, undoubtedly as a direct result of his chemical experiments. The heating of mercury oxide and working with hydrogen cyanide in an unventilated space are not the most healthy occupations.

Priestley, on the other hand, was penalized for his open support of the revolution, even though he lived on the other side of the Channel. His arguments in favour of science and rationality struck many as outrageous, especially coming from a theologian. Despite his scientific merits, he was forced to flee to the USA in 1791 after an angry mob set fire to his church and house, thereby reducing both to ashes and $CO_2$. Fate seemed generally to frown on the trio who discovered $CO_2$, but at least Priestley ended his days in 1804 – not under the guillotine, but at the age of seventy in Pennsylvania.

Around the same time, another of Sweden's renowned scientists, Jöns Jacob Berzelius,[15] gave a formal name to the subject of this book. The transition from the 1700s to the 1800s was a revolutionary period for chemistry in general and carbon in particular. It was Berzelius, however, who transformed carbon to C. As his more famous countryman, Carl Linnaeus, accomplished within biology a few years earlier, establishing the binomial nomenclature for every living thing, such as *Homo sapiens* and *Rattus norvegicus*, Berzelius discovered basic elements and identified them with simple letters. It was a taxonomical process inspired, in all likelihood, by Linnaeus. Some elements required two letters: since C was already allocated to carbon, for example, calcium was termed Ca. Berzelius also introduced the modern designation for the number of atoms in compounds, such that $H_2O$ continues to denote that water contains two hydrogen atoms. The only difference was that Berzelius wrote $H^2O$.

Berzelius also suggested that elements could be divided into organic and inorganic groups, with almost all carbon compounds being located in the organic group, whereas salts, metals and water were inorganic. Chemistry has retained this division and every chemistry student takes various courses in organic and inorganic chemistry, with organic chemistry being largely related to the living world, though there is no sharp distinction. Organic chemistry, in short, tackles the compounds containing C and H (hydrogen), whereas inorganic chemistry encompasses everything else. Along with Robert Boyle, John Dalton and Antoine Lavoisier, Berzelius is part of the quartet we consider to be modern chemistry's founders. Personally, I also credit Berzelius for the launch of stoichiometry, but more on that later.

Of course, it was not only Berzelius who wanted to systemize things and create a general overview. The Russian-born chemist Friedrich Konrad Beilstein began studying chemistry at the University of Heidelberg at the age of fifteen.[16] One of his ambitions, which he believed could be achieved, was to assemble a list of all the known carbon compounds. In 1881–3 he published the first edition of the work in two volumes, comprising 2,200 pages and 15,000 compounds. Already prior to publication, however, he realized that this subject would call for endless revision.

The second edition appeared in 1886 and consisted of three even bigger volumes. When the third edition appeared in 1900, this time in four oversized volumes, it was clear the subject was in danger of exceeding the limits of paper publication. Here we can again draw a parallel to Carl Linnaeus, who thought he was close to cataloguing the world's assortment of species when by the 1770s he had succeeded in naming an admirable 8,000 plants and 6,000 animals, and had concluded that the earth's total diversity amounted to 26,500 species.[17] Today's projections swing from between 2 and 10 million species, but we will never know the true number. Unlike in the world of organic chemistry, we have not yet reached a point where new species are being created in the lab. However, it may be only a matter of time. The first synthetic bacteria have already seen the light of day, and the most visionary souls believe that 'soon' we will reach a point where most carbon-rich products can be created from synthetic organisms, and that these will fulfil our need for biofuels even as they allow us to get a handle on $CO_2$.

Today chemical mapping takes place through the Beilstein database, the world's largest database of organic chemistry, and the *Beilstein Journal of Organic Chemistry* is one of the places where new compounds and chemical reactions are continually published. It has been estimated that the number of carbon compounds currently equals the number of species on earth: 10 million. I will *not* address all of them here. Even if carbon atoms so enjoy each other's company that up to sixty can bind exclusively together without inviting other elements into the mix, no one can accuse carbon atoms of lacking friends. It is easier to name the combinations that do *not* contain carbon than those that do.

## THE MASTER OF DISGUISE: SOFT, HARD AND ROUND

When carbon binds with hydrogen a transition occurs from inorganic to organic chemistry.[18] Chemistry makes a basic distinction between inorganic and organic chemistry, which essentially assumes that compounds in which C binds with H are typical for living material. The simplest organic molecule is methane, $CH_4$, which is produced by life but which is not a part of the organism. Indeed, a large quantity of the almost endless number of organic compounds are synthetic.

It is worth noting that *organic* has a different connotation here from those it assumes in other contexts. For example, the insecticide DDT (dichloro-diphenyl-trichloroethane), a molecule consisting of two hexagons with carbon in the corners and five chlorine atoms attached, is chemically considered organic, whereas in other contexts food grown in the absence of insecticides like DDT are considered organic. In this case chemists could be said to have priority on the term organic by invoking a 120-year-old tradition. Furthermore, whereas the formula for DDT is $Cl_5H_9C_{14}$ – that is, five chlorine atoms, nine hydrogen atoms and fourteen carbon atoms – the structural formula for DDT is usually written like this:

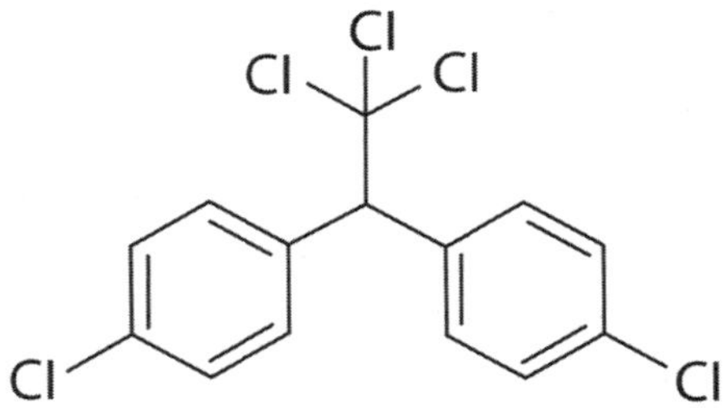

DDT – an organic molecule, although banned when growing organic food. A carbon atom sits in each corner of the hexagons.

Here it is taken for granted that every corner and bond holds a carbon atom (there are fourteen carbon atoms in total). This fact is so obvious within organic chemistry that the letters do not even appear in the formula. Just as obvious are the nine invisible H atoms that form the fourth bond for every carbon atom, as well as the double lines symbolizing a double bond (remember that carbon has four electrons and therefore normally four bonds).

Generally speaking, carbon assumes three pure forms in nature: graphite and diamond, both of which have long been recognized, and that bizarre newcomer, fullerene. We all know that pencil 'lead' and diamonds are both pure carbon, and we are fascinated by the fact that what can be extremely soft and cheap material can also be very hard and expensive. One would think the idea that diamonds did not just belong to the carbon world, but were also pure carbon themselves, would be a relatively new discovery, given how counter-intuitive it is that hard, pure diamonds would be formed from the same material as wood charcoal and pencil lead (which is, of course, graphite). Actually, Lavoisier had already realized in 1772 that diamonds were carbon. Not too long after, the discovery was also made that pencil lead was not lead but carbon: during combustion both diamonds and graphite give off an amount of $CO_2$ equal to their weight.

The most famous diamond in world history, and one of the largest, is the Koh-i-Noor. Originally 793 carats, it is said to have been following the world's great rulers for 5,000 years.[19] The largest recorded diamond, the Cullinan, was found in South Africa in 1905 and was around 3,106 carats (620 g) uncut. Our enduring fascination with diamonds is undoubtedly due to their rarity and durability, but even diamonds do not necessarily last forever – one can, in fact, burn them. That said, their melting point is 3,200°C and their boiling point 4,200°C, but they also freely burn at temperatures over 800°C according to the simple reaction equation of $C + O_2 \rightarrow CO_2$. Yet, how can the same material form the basis for these extremes? Even more astonishingly, graphite can be transformed into diamond and diamond into graphite 'simply' by applying the correct combination of pressure and temperature.

The oldest known graphite also perhaps provides evidence of the oldest known planetary life. This graphite is found in a deposit of 3.8 billion-year-old sedimentary rock in Greenland, and is so pure you can write with it. It contains the isotope 13C (to which we will return), which strongly suggests a biological origin, though we will never be absolutely sure. Graphite itself consists of layer upon layer of six carbon atoms connected hexagonally in what looks to be a chicken wire net of pure carbon, and the layers are bound together by weak forces, so that the layers divide easily, for example, when the pencil lead meets a little friction on paper. Diamonds, in contrast, feature carbon atoms bound together in an almost indestructible network, forming one of the hardest materials known to man.

Diamonds are formed under extreme pressure and temperatures, and naturally come from deep inside the earth's pressure chamber and baking oven. Synthetic diamonds are produced by replicating these processes, where a carbon plasma cloud is subjected to high pressure and temperature in the laboratory. Chemically these are true diamonds, but since they are 'artificially' rather than 'naturally' produced, they have a much lower value, at least on the gem market. It is always interesting to note the precedence something 'natural' takes over something 'artificial',

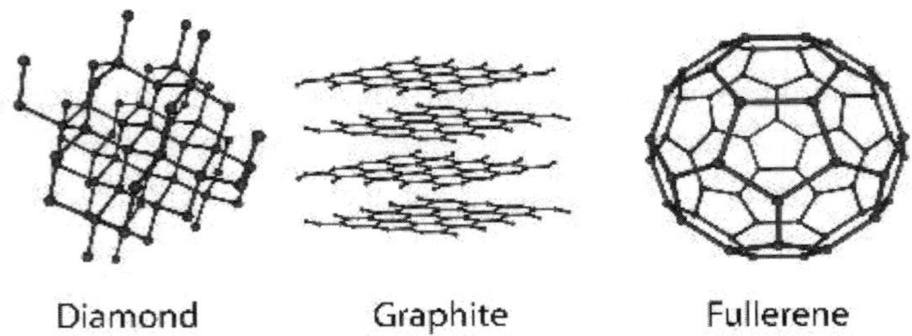


Diamond, graphite and fullerene are the three main forms of pure carbon. While diamonds obtain their unique hardiness due to their crystal structure, with covalent bonds between each carbon atom to four other atoms, graphite has a layered and 'slippery' structure. Fullerene represents a third and completely different structure by creating a football with 60 carbon atoms.

since what is natural is perceived as being authentic, whatever the chemical truth of the matter. Perhaps it is only within the pharmaceutical industry, where many of nature's intricate carbon compounds, those yielding medicinal effects, have been copied with great success, that we fully value synthetic products.

Who first created artificial diamonds is a matter of dispute, but in 1955 the General Electric Company succeeded in industrially producing artificial diamonds on a large scale, and the company's value increased by more than $300 million in one day.[20] Today the annual production of synthetic diamonds, which are mainly used for industrial purposes, totals more than 100 tons.

MORE THAN 200 years after Lavoisier, however, it turned out that carbon had more surprises to offer. Diamond and graphite were not the only pure carbon forms found in nature. In 1985 another fundamentally different form of pure carbon was discovered: fullerene. That curious name was originally even stranger, since it was first called Buckminsterfullerene, named after the architect Richard Buckminster Fuller (1895–1983), who constructed fantastical domes composed of pentagons and hexagons.[21] As there are limits on the number of times one can pronounce and write Buckministerfullerene, fullerene became the compound's working name.

The discovery of fullerene was the unexpected by-product of an entirely different enterprise: the search for long carbon chains formed in the stars. In a complicated experiment, the processes at work within gas planets were replicated. The result of this experiment yielded something strange: molecules that consisted of exactly sixty carbon atoms – a chemical chimera, unlike anything known in the world of carbon chemistry. Carbon structures can assume many different forms with all varieties and numbers of carbon atoms: according to many people, the fact that so many of the molecules turned out to have exactly sixty atoms could only mean that this was an entirely unique structure, a kind of closed unit (which is not something one can simply put under a microscope to study).

Many variants of the experiment were repeated, but the mystical sixty carbon molecules continued to appear. The only logical explanation for sixty pure carbon atoms was a molecule composed of twelve pentagons and twenty hexagons. Harold Kroto and his colleagues at Rice University in Texas described the structure as a polygon with sixty edges and 32 surfaces – a soccer ball. The image of the football 'on Texas grass' became iconic among chemists. The title of Kroto's article also became iconic with its two words: '60-C: Buckminsterfullerene'.[22] The journal *Nature* cleared its front page in honour of the subject and the iconic carbon football decorated the November 1985 issue.

By a twist of fate, the ultimate man-made fullerene dome was also built in the Arizona desert around the same time. Of course, the drawings and plans already existed before the fullerene atom appeared on the cover of *Nature*, but the actual construction commenced in 1987. We are talking here about Biosphere 2, an attempt to create an intact, self-sustaining, closed ecosystem in a glass dome.[23] The structure was formidable in every sense of the word and contained 1,900 $m^2$ of rainforest, 820 $m^2$ of ocean with a coral reef, 1,300 $m^2$ of savannah, 1,400 $m^2$ of fog desert and 2,500 $m^2$ of arable land. It was based on an intricate circulation system of water, air and solar energy coming through the glass. On 26 September 1991 the door closed behind four men and four women who would spend the next two years inside the bubble. Strictly speaking, it was also a social experiment that worked astoundingly well, both sociologically and biologically. The eight people managed fine with what they cultivated and harvested inside their glass house, avoiding the worst conflicts. The carbon cycle, however, proved far more difficult to manage.

Whereas Biosphere 1 (that is, the earth) has maintained remarkably constant $CO_2$ and $O_2$ concentrations over many hundreds of thousands of years, within Biosphere 2's closed ecosystem the gases oscillated like a yoyo. Daily swings in $CO_2$ could exceed 600 ppm (parts per million), with low values occurring during the day (as plants assimilated $CO_2$) and high at night. Winter was even worse when plant production reached its low point and $CO_2$ increased toward 4,500 ppm, almost twelve times the level on the outside. Of course, all this took place in a greenhouse, with a ventilation system, so the $CO_2$ did not contribute to any warming, nor is $CO_2$ toxic.

The more serious concern was that the eight individuals in the fullerene dome fell victim to an increasing lethargy. This state was not due to high $CO_2$ levels, but to a fall in oxygen levels, from 21 per cent to 14.5 per cent. That was equivalent to living at an elevation of 4,000 m, where thoughts and movements become sluggish. One explanation was the significant amount of carbon-rich earth that emitted $CO_2$ while consuming $O_2$ through bacterial decay of organic matter. There was also a large quantity of calcium-rich concrete in the foundation that accumulated $CO_2$ according to

Biosphere 2 in Arizona provided important insights into the $CO_2$ cycle.

an equation we shall encounter numerous times: $Ca(OH)_2 + CO_2 \rightarrow H_2O$. This was not enough to much affect the sky's high $CO_2$ level, but was enough to ensure a perpetual drain of $O_2$ bound into $CO_2$. On top of that, there was also an increase in the third main greenhouse gas, nitrous oxide ($N_2O$), to nearly life-threatening levels, which also altered the internal nitrogen cycle. Ultimately, in order for the project to continue, they were forced to cheat by pumping in fresh air. Nature is not easy to imitate, but if nothing else, Biosphere 2 was an impressive attempt.

Some years after my own humble studies of the carbon cycle of the lake in a Norwegian spruce forest, I visited Biosphere 2. In many ways the Arizona desert presented a stark contrast to the wet, carbon-rich system in the forest where trees, black forest soil and a swamp several metres deep are all formidable carbon repositories in a cold and wet climate. The Arizona desert, like most deserts, is hot, dry and extremely carbon-poor. Sand does not hold much carbon, and the thrifty cactuses and creosote bushes surrender almost all of their stored $CO_2$ back into the atmosphere when they break down in the dry climate. Therefore, one of the most striking things about Biosphere 2, with its astounding micro-forests, micro-coral reef, humidity and lushness, was the contrast it made to the sunburned landscape outside.

The dome with its small ecosystems is still intact and is well worth a visit. When the glass dome is finally removed, it will not take many years for all the life-giving carbon sheltered beneath the dome to oxidize in the baking sun and disappear, along with the water, back into the atmosphere. In many ways this is a dark metaphor for the worst case scenario in store for Biosphere 1 – earth – only in this case it is the greenhouse effect we must fear. This has perhaps been a long desert trek from fullerene, but Biosphere 2 is a useful reminder of why we need to value a balanced carbon cycle and watch the amounts of oxygen, methane and nitrous oxide, none of which we can take for granted in the Biosphere's original version.

WE HAVE seen that the extremes of graphite and diamond are the main forms of pure carbon in nature, and we can now add fullerene as a third, much rarer form. The number of pure carbon compounds is also increasing, but the actual number of pure carbon compounds, combinations and eventually synthetic material, where carbon establishes bonds with others, is unimaginable. Carbon is not just life's essential molecule, but the raw material for an endless array of new, high-tech products. In the realm of carbon products alone, it is possible to create C70 fullerene, and nano tubes and single-layer carbon (graphene) have opened up a wealth of potential uses.[24] Some scientific discoveries are the fruits of long-term, targeted efforts; others are the result of failed attempts or pure coincidence.

The dream of a one-atom-thick carbon sheet was certainly not new, but when that dream was finally realized it happened rather unexpectedly late in 2004, when Andre Geim and Konstantin Novoselov, in a fairly simple process, isolated a single carbon layer from a bit of coal on a piece of tape, which earned them the Nobel Prize for Physics in 2010. This ultra-thin carbon layer, known as graphene, which consisted of the familiar hexagons, as if someone had cut an ultra-thin cross-section of a beehive, harbours astonishing characteristics in terms of tensile strength and conductivity. It is three hundred times harder than steel, almost transparent, an excellent electronic conductor, elastic and non-combustible.

Carbon fibre also yields a combination of strength and lightness with which no ordinary metal can compete. For example, it is far superior to steel and, like graphene, seems to present endless possibilities. Some have imagined that single-layer carbon coupled with DNA will result in DNA-based chips that will revolutionize data storage and become the ultimate synthesis of life and non-life. DNA is, after all, a digital code containing unimaginable quantities of information, all based on a simple, four-letter alphabet, meaning all the information necessary to create a human resides in a microscopic cellular nucleus. All the variety of possibilities necessary for forming 7 billion unique individuals is reached by varying the order of these four letters, and the quantities of information contained in the DNA of a single microscopic nucleus is enough to fill 6,000 books, each with five hundred pages. That is an endlessly more effective form of information storage than even the smallest silicon chips in a supercomputer, even employing a single atom carbon sheet as the foundation. Who knows? Many of our most brilliant discoveries are imitations of solutions and designs that evolution produced millions of years before our time.

Carbon tubes, which can be used to store hydrogen gas in gas-driven motors, are also high on the list of promising materials. They are sensitive detectors and among other features are useful within the pharmaceutical industry. So far we have only considered pure carbon materials. When carbon bonds with other materials, that is when everything really starts. As mentioned, there are an estimated 10 million different carbon compounds, and new ones are constantly emerging in laboratories the world over. Carbon certainly has more surprises up its sleeve in the years ahead.

## NYLON TIGHTS AND THEIR DESCENDANTS

Carbon's success as a chemical collaborator is partially due to the fact that it can shift between single, double and triple bonds, not to mention combinations of these, as needed, as long as all four electrons find a suitable partner. In carbonic acid, $H_2CO_3$, carbon forms a double bond with the free oxygen atom and a single bond with the two others, since these also have a single bond to hydrogen. In $CO_2$ there is naturally a double bond between C and each of the oxygen atoms. One might assume that carbon monoxide (CO) has a quadruple bond, but instead there is a triple bond

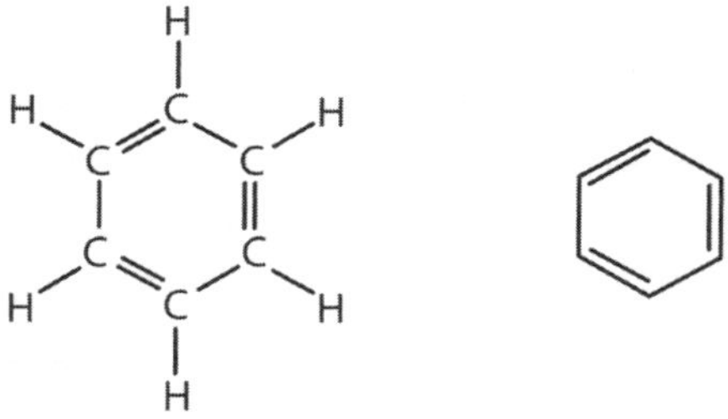

Benzene – the version on the right for those who have studied the matter.

between C and O, something that gives the molecule a weak charge and contributes, among other things, to the fact that CO is unfortunately effective in binding to the haemoglobin in red blood cells. The fact that CO outcompetes $O_2$ in blood cells is the reason that CO, unlike $CO_2$, is toxic.

Carbon's favourite structure, however, seems to be a ring, something that we have seen in graphite, graphene and fullerene, not to mention in organic structures such as DDT and millions of others. Such carbon rings are central to life, as well as to countless synthetic components, primarily in the pharmaceutical industry. Around 75 per cent of all organic molecules contain ring structures in some form or other – often different forms together. When they occur in such family groups of six and six, they generally have a double bond to one neighbour and a single bond to another, and when it comes to their fourth, free bond, imagination is the only limit. One of the simplest molecules within this enormous genus is benzene, first described by August Kekulé in 1865.[25]

Even if the molecule were simple, the discovery was huge, and it was no accident that Kekulé himself 'discovered' it, or rather, deduced it according to his theory regarding how electrons in the outer electron shell would logically form bonds to each other. What we today call 'bonds', of course, he, with typical German affinity for plain terms, called *Verwandtschaftseinheiten* (relationship unity). Kekulé was among the converts, those who have departed their original profession after experiencing a divine calling. In this case, it was Justus von Liebig who convinced Kekulé to abandon architecture for chemistry. Liebig, it might be added, also taught Friedrich Konrad Beilstein, the man who set out to classify all organic chemical compounds.

The material itself had already been extracted from coal tar in 1825 by Michael Faraday. Later it was given the name benzene. (The benzine with which we fill our tank is composed of variously branched hydrocarbons with between four and twelve carbon atoms.) It was clear that behind benzene's characteristic, almost sweet smell stood $C_6H_6$. But how all these Cs and Hs were connected was a mystery. Unlike the much later fullerene, there was no logical structure that could yield a stable relationship. At least not until the Lord of the Rings, Kekulé, then a professor at Ghent in Belgium, apparently dreamed a solution in which the molecules twined together to form a snake biting itself in the neck – the ring structure. Eureka! Kekulé's story has a few weaknesses, according to those who have examined the matter more closely; nonetheless, Kekulé described the first of a number of aromatic structures that have as a common denominator the benzene ring, and its relatives that apparently form the basis for three-quarters of all organic chemistry.

Organic chemistry might sound peculiar to those unacquainted with the subject, and should perhaps be taken in large doses only by those with a special interest. Nonetheless, it harbours an amazing logic and symmetry, while at the same time relating to most of what is happening around and inside us. Once you have mastered a few basic principles within organic chemistry – and here carbon's bonding properties and affinity for hexagons are central – much falls into a logical pattern. This is not a book on organic chemistry, but before we follow the carbon atom into the carbon cycle's labyrinth, we ought to glance further into the world of chemistry.

Organic molecules fall into the main groups of hydrocarbons, aromatic and cyclic compounds (benzene's countless relatives), alcohols (more than you would expect) and heterocyclic compounds. The basic logic within carbon chemistry is simple and follows a clear system.[26] Hydrocarbons are the easiest and most comprehensible group; one just needs to keep track of C and H. The simplest of these are methane ($CH_4$), ethane ($C_2H_6$), propane ($C_3H_8$) and butane ($C_4H_{10}$) – all flammable gases and with single bonds between the Cs. Next come the compounds whose name indicates the number of carbon atoms they contain: pentane (5), hexane (6), heptane (7), octane (8), and so on. This is followed by the alkenes, very similar in name and structure, but with a double bond between C atoms, and consequently less room to accommodate H atoms (such as ethene – $C_2H_4$, propene and butene). And the next group, alkynes? As expected they have a triple bond between Cs and, naturally, the first in line is ethyne ($C_2H_2$), better known as acetylene. These three groups are the aliphatic compounds.

Aromatics, as mentioned, are ring compounds, such as benzene, toluene and naphthalene, all with C in a hexagon or a pentagon in every combination and with 'tails' of other chemical compounds in an endless number of combinations. They often have a characteristic smell, which children often find pleasing, whereas adults have learned that they are toxic and dangerous. Benzene, for its part, has the unmistakable smell of organic chemistry. Name-wise, alcohols follow the nomenclature from methane to methanol and so on, and the common denominator here is that an H is exchanged with an OH: ethane ($C_2H_6$), for example, becomes ethanol ($C_2H_5OH$). In any case, this just scratches the surface or, rather, is a quick peek at the most basic of building blocks that can be tacked together to make virtually anything.

Immediately after the First World War chemists began to splice and combine organic molecules in ways nature itself had not had the vision, reason or possibility to produce. First in line was Hermann Staudinger.[27] He became interested in polymers, which simply means many (poly) parts (mers); accordingly, they can be used for almost anything composed of multiple parts. Staudinger began with polymers found in natural materials and, to his credit, showed that apparently large and complex molecules, such as those that formed starches, proteins, cellulose or rubber, actually consist of long chains of small molecules connected by covalent bonds, like a pearl necklace.

Although Staudinger and his colleagues created *new* polymers for new purposes, polymers themselves were by no means new to the natural world. As with almost everything else, evolution over hundreds of millions of years had already discovered polymers to be a robust structure, for example, in the form of cellulose, which is a polymer of glucose or proteins (a repetition of twenty amino acids), but in the lab new material could be connected to create new polymers with astonishing properties. Many of these materials formed long, durable and lightweight textile fibres, such as nylon and polyester. Nylon, for its part, saw the light of day at DuPont's laboratories on 28 February 1935 and launched an entire industry with carbon chemistry in a starring role.[28] Nylon stockings, toothbrushes, parachutes – a whole new market was opened for synthetics. DuPont itself started small with gunpowder production in 1802, but quickly realized that other branches of chemistry offered great possibilities. In 1927 they expanded their research budget fifteen times over, and the effort bore fruit. Today it is among the world's largest chemical companies.

Central to DuPont's success was a man fifteen years Staudinger's junior, Wallace Carothers (1896–1937), who was closely involved in the development of nylon, as well as neoprene and polyester. Neoprene was the first completely synthetic rubber and is familiar to all divers and surfers who have ever used a wetsuit, whereas polyester consists of chains of esters. Carothers did not bask in the glow of his success, but rather believed that, despite the success, he had failed. On 28 April

1937 he checked into a hotel in Philadelphia and mixed himself a cocktail of lemon juice and potassium cyanide – KCN, carbon again, this time in bad company. The bond between calcium and nitrogen forms a crystalline salt that looks like sugar but is extremely toxic. That was the end for Carothers, but the polymer adventure continued.

It is high time to correct the heretofore regrettable gender disparity in the history of carbon. Stephanie Kwolek started at DuPont after the Second World War, also on a quest to find new materials in nylon's wake. She began with a six-carbon ring that had an amine and a carboxylic acid as its side branches. This compound could only be dissolved in sulphuric acid, which reduced it to a porridge-like mush that, stated rather simply, produced an ultra-strong fibre after being pressed through an extremely thin hole. This process was simplified and refined many times until it yielded a product that could be mass produced and woven together to form Kevlar. Kevlar is lighter but five times stronger than steel, and replaces steel in bulletproof vests and helmets, as well as in sails, cables, guy ropes and other products that require a combination of resilience and light weight. It also tolerates temperatures up to 400°c. Perhaps graphite, diamond and Kevlar represent carbon's extremes.

Before we get entirely carried away by our human intelligence, however, it is worth remembering that nature beat us to the punch here. The polymers in spider silk consist of chains of large proteins whose tensile strength relative to their thickness exceeds even Kevlar. In addition, the threads have an antimicrobial effect, but are nonetheless completely biodegradable. As such, we are simply following in evolution's tracks when it comes to applying smart solutions within the realms of design or medicine. Indeed, significant research is now being devoted to reproducing spider polymer chemistry on a large scale.

## CARBON ON WHEELS

When the sun had driven away the fog after the night's rain up on my lake, I rowed out to retrieve the bottles filled with brown, carbon-rich lake water, which had been treated with radioactive carbon in order to measure algae's photosynthesis. As the algae worked, we shook the water from the tent and packed up the microscopes and our other experimental odds and ends from the miserable and makeshift lab beneath the spruce trees. I retrieved the bottles, which are made of quartz, though the rest of the equipment is mostly plastic and Bakelite, polymer chemistry's fruits. We stacked the cardboard boxes – cellulose, another polymer chemistry product – and packed it all into the university's beaten up Volkswagen Caravelle. And where would that automobile have been without carbon and polymer chemistry?

In 2010 the world surpassed a total of 1 billion cars: 60 million new ones are produced annually. Just a century before cars were scarce, even though Henry Ford began production of his Model T in 1905.[29] Well, *some* cars did exist. A steam-powered car was built in 1769, but it remained a prototype. That was also largely the case for its successors. A hydrogen-powered car rolled a few metres in 1807, but it produced no direct successors. The developments that enabled the evolution of the modern car were perhaps Nikolaus Otto's four-stroke engine or Karl Benz's three-stroke engine. Both were based on transforming hydrocarbons to mechanical energy through combustion. However, it was Henry Ford who really founded the industry. The car's essence, furthermore, was benzine, or diesel, applied by Rudolf Diesel in his first four-stroke engine.

The wheel is also clearly a central component of a motor vehicle, and the wheel was indeed reinvented.[30] After the compact wooden wheel emerged some 4,000 years ago (presumably before that time pyramid builders and others used comparable techniques that allowed them to move tons of rock by dragging them over round logs), there came the significantly lighter spoked wheel, then the metal-encased wheel, before the rubber wheel finally emerged in the late nineteenth century. First came the compact rubber wheel, which must have been like driving on a bouncy ball rolled onto a wooden wheel, and not especially comfortable given the day's uneven surfaces. Then came the inflatable rubber wheel at the end of the 1880s, the invention of which was, strictly speaking, a dead heat between Britain's John Dunlop and France's Michelin brothers. In both cases a bicycle was involved: John Dunlop equipped his son's tricycle with inflatable rubber wheels, whereas in France the scenario, naturally enough, involved a bicycle racer. Eduard Michelin was behind the first interchangeable and, eventually, mass-produced bicycle tyre, and the prototype was apparently fitted on the bicycle that won the first Paris–Brest–Paris race in 1891, twelve years before the first Tour de France.

Whether originating with Dunlop or Michelin, the air-filled tyre came as a relief to road travellers, whether going by car or by bicycle, not to mention by horse-drawn carriage, which continued to be the dominant mode of transportation. Rubber is nonetheless the keyword, and synthetic rubber in its various forms consists of polymers of petroleum-based monomers hooked together to form long, elastic chains.

A car tyre contains yet more carbon in the form of soot. Over 8 million tons of soot are produced annually by the incomplete combustion of coal and petroleum products, and 70 per cent of this ends up in car tyres (which are two-thirds soot), contributing to the tyres' colour and durability. Soot, also known as 'black carbon', is often used as a pigment in other areas. Battery components, the car's plastic body and most of its interior also originate in products whose central feature is carbon.

When Henry Ford started mass production of the Ford car, the sky was bright and it never crossed anyone's mind that this foremost symbol of human progress and freedom would also have a downside. We know that now, of course, and putting aside traffic jams, accidents and the steadily increasing need for asphalt, it is obvious that $CO_2$, the by-product of combustion, is the problem. The Michelin Man has become the symbol not just of car tyre production, but of the overweight human with a spare tyre around their waist – in large part thanks to the combination of too much carbon, false carbon, and the fact that cars made it possible to transfer the energy used in movement from the body to the car and benzine.

Synthetic rubber has any number of external uses, but other types of artificial carbon are to an increasing extent used internally. Wine gums, jelly candies and other foods contain large quantities of carbon chemistry and sweeteners like glycerol ($C_3H_8O_3$), or sugar alcohols (which, it must be said, only give you a sugar rush) like sorbitol ($C_6H_{14}O_6$) and other related compounds, mannitol, lactitol, xylitol, all of which are structurally similar to common sugar ($C_6H_{12}O_6$), but have a more intense sweetness. All these compounds, furthermore, have the same energy content as common sugar, whereas energy-poor sweeteners such as cyclamate or aspartame (used, for example, in soda drinks) have a sweetening effect that is sky-high compared to that of sugar. Whether these sweet fruits of carbon chemistry are worse than sugar simply because they are 'artificial' or 'synthetic' is a matter of debate, but they do not rate a Michelin star.

## PLASTIC FANTASTIC

Long before iron took over, our culture was based on carbon from rocks and trees. We are now in the plastic society. Plastics are also polymers, such as those we find in nylon, rubber and many other things we consider to be carbon's synthetic products, but to plastics are usually added various curing agents.[31] When plastic first fully entered the marketplace, making its way into kitchens, industry and daily life, it was considered to be one of the very signs of progress. The old, heavy ladles and wooden bowls were out, everything literally became lighter, and since that time plastic has conquered our everyday lives in ways even the most visionary polymer pioneer could not have dreamed of.

By 1908 the cured plastic material Bakelite was already on the scene. The name stems from its founder, Leo Bakeland, not because the material has been slightly baked. In the 1930s Bakelite was followed by other synthetic plastic materials such as polyethylene, polystyrene, polyvinylchloride – and also nylon. Plastic made its way into not only kitchen cabinets, but insulation material, hoses, bottles, glasses, furniture, flooring, toys, clothes, indeed almost anything. Plastic was modernity itself, proof that the world was progressing. Oil, plastic, indeed synthetic chemistry itself demonstrated how human cleverness could further develop the ingenuity already displayed by nature when it came to finding uses for carbon.

The surge of excitement surrounding modernity and progress faltered a little during the 1970s when it turned out that there was another side to plastic's development. Even plastic became suspect. In the first place, it symbolized something 'fake', in contrast to nature's raw material. Whereas the wooden ladle in the kitchen drawer was 'real', plastic ladles, if not exactly 'fake', were 'synthetic', which again had the aftertaste of something false. Plastic is obviously an artificial material, not especially environmentally friendly either, but the most important objection for many here was that it was not *genuine*. And the idea of natural (as something that equals nature) has the same appeal as something genuine; owners of wooden boats, for example, tend to claim that their boats have a 'soul'. Yet paradoxically, forests have now come to form the springboard for increasingly more synthetic polymers, hydrocarbons, plastics and other things with an origin in cellulose and lignin.

Plastics, and especially plastic bags, have also come to symbolize human waste.[32] Although plastic sooner or later does eventually succumb to bacteria, fungi and sunlight, it is later rather than sooner. No organism is especially effective at breaking down plastic. Oil, on the other hand, gets broken down rather quickly since it is 'natural': in many ecosystems where sufficient oil leaks from the depths, evolution has produced bacteria that know how to exploit this energy source. Plastic, in contrast, is tough fare, and especially hard plastics. Even at 80 degrees north, on the northernmost tip of the Svalbard Islands located halfway between Norway and the North Pole, plastic washes ashore. During the summer of 2014 I travelled to the Svalbard Islands on a research project. Even here in the remote, high Arctic, I waded through colourful plastic on the beaches: Russian ketchup bottles, British plastic boxes, Norwegian plastic bags and a random assortment of other plastic products, floats, fishing nets and nylon rope of unknown origin. The distinction between genuine and 'plastic', of course, is also not entirely clear: oil is as natural as the wood it comes from, and there should be no natural law compelling us to slap a negative label on everything that is processed (or, in this case, let us say 'refined') by us. Nonetheless, an orange plastic container on an Arctic beach does not feel as natural as driftwood. Plastic is quite simply everywhere.

It is estimated that somewhere between 500 billion and 1 trillion ($10^{12}$) plastic bags are produced annually. Where do they all go? Fortunately, some are recycled, but many end up in landfill, on beaches and in the oceans. When a 'plastic ocean' of drifting plastic fragments, covering an area larger than Norway, was discovered in the Pacific Ocean in 1999, alarm bells began to ring. The plastic ocean, granted, is not entirely visible, being composed mainly of small plastic fragments. Not that that improves the matter. In 2009 the UN estimated that annually 6.4 million tons of garbage ends up in the ocean, and that today there are 100 million tons of waste in the world's oceans, the lion's share of that being plastic. An estimate from 2015 suggests that the world's coastal countries generate a total of 275 million tons of plastic waste annually, and that between 5 and 12 million tons of that ends up in the ocean.[33] Even if the plastic bags do eventually fragment that does not solve the problem, in any case not for incoming generations of seabirds who continue to ingest plastic fragments, either directly or through the food chain.

As a boy I was able to talk my way onto a trip to the bird island Runde. It was actually an older cousin who had received the invitation to sail aboard the Norwegian newspaper *Sunnmørsposten*'s boat *Svint*, but I simply *had* to go. It was a powerful experience in every sense to stare up at the countless flocks of birds, but what I best remember was the *sound* – the overwhelming orchestra of more than 100,000 breeding pairs of kittiwakes like a snowstorm against the cliff walls. I have since returned several times and have always been met with this astounding impression of life. In 2014, however, I was on Runde again before Easter for a conference concerning what was happening to the ocean and to the birds. We travelled around the island – and it was silent. Hardly a kittiwake, hardly an auk. Where 5,000 fulmars once nested, it was now dead. Skarveura (the shag rockery), once the world's largest colony of common shags, a kind of cormorant, was also cold and dead. Later that evening we climbed to the top of the puffins' nesting sites and to my relief the puffins were still there.

The herring population rebounded just in the nick of time for the puffins. Puffins were the ocean's canaries. Just as miners used canaries as an early warning sign of dangerous concentrations of methane gas, even before the miners had detected them, so starved puffin chicks had warned us that something was fundamentally wrong deep in the ocean – we had fished the world's largest fish stocks to nearly nothing. Gulls, fulmars, guillemots and cormorants are also the ocean's canaries, warning us of other types of changes we cannot see down in the depths.

It is a silent spring on bird mountain, where only the gannets persist.[34] They have entered the new age and in many places now build colourful nests of nylon rope, only the change-resistant maintaining their 'genuine' old nests of seaweed. What is happening to the seabirds is due to a complex history of both overfishing and climate change. Even if seabirds have always experienced population fluctuations, the collective decline is alarming. Researchers have also identified a new suspect – plastic. Oil spills have long been regarded as a primary threat to sea birds, and with good reason. However, it seems oil is just as detrimental to birdlife when it occurs in the form of plastic.

One finds heart-wrenching scenes on YouTube of dead and dying young albatrosses with stomachs full of plastic. Fulmars, storm petrels that sail over the sea surface in search of plankton and other food, fill their stomachs largely with plastic, and that might be one of the reasons the species has suffered such dramatic declines across the globe. One study reported finding plastic fragments in 95 per cent of the fulmar stomachs examined. One of the places fulmars hunt for food is the North Sea, which also has its share of plastic. Annually, 20,000 tons of rubbish are dumped in the North Sea, and about three-quarters of this is plastic. Unfortunately for fulmars, around 15 per cent of this remains floating, and unfortunately for a number of benthic organisms, 70 per cent remains on the seabed. The plastics that overrun beaches account for only 15 per cent of the total. On top of this are the actual microplastics, the micrometre-sized spheres that are the new 'must have' abrasives in toothpastes and skin creams.

A number of other species also ingest plastic, and even when plastic fragments break down into microparticles it enters the food chain. Plastic also contains a number of organic environmental toxins such as PCB, PAH, pesticides, flame retardants and other organic, carbon-rich molecules. Various kinds of soft plastic are particularly notorious for containing phthalates, since esters of phthalic acid have the unfortunate and unintended effect of functioning as endocrine disrupters. The body mistakes them for oestrogen, and in the worst cases it leads to gender confusion and impaired reproductive ability. Even the top levels of the food-web are affected. An adult male Cuvier's beaked whale found stranded on a beach in southwestern Norway in early 2017 was in bad shape. It turned out that the poor whale was filled with plastic, with no fewer than thirty plastic bags clogging his entire gut and digestive system. His poor fate did, however, open many people's eyes to the alarming plastic pollution of the marine ecosystems.

When Alan Weisman wrote his prizewinning bestseller *The World Without Us* in 2007 he underscored how quickly the traces of our lives will rust, crumble and become overgrown.[35] One thing, however, will survive as the last evidence of our civilization thousands of years after we have vanished: plastic. I do not know if Weisman is right. I imagine, for example, that the pyramids, which were here 4,000 years before the first plastics, will remain 4,000 years after plastic disappears, at least as ruins, but the point is still clear. It takes an estimated twenty years for a plastic bag to break down, but that is irrelevant when it is continuously being replaced by new ones. Plastic bottles and disposable diapers take about four to five hundred years to break down, and a nylon rope that was dumped in 2000 will remain until 2500. If you outfit, in other words, a wooden boat with a nylon rope, the rope will disintegrate long after the boat and the pier have returned to the atmosphere as the $CO_2$ from which they came.

With this grim glance at the disadvantages of plastic's durability, we will leave the world of synthetic polymers without casting any obvious moral judgement. Through polymer chemistry, carbon has certainly given us products with amazing characteristics, and our world today would be unimaginable without them. Plastic is just one of the countless examples of too much of a good thing, which perhaps just reflects an overabundance of people, or at least an overabundance of people fiddling with carbon in too many ways.

## SYNTHIAS AS A CARBON COMPANY

In many ways, our human project has been to achieve mastery over nature. Though for much of our prehistory we have been subject to nature's whims, we have gradually come to understand various causalities and have since used these to tame fire, animals, waterfalls and disease. At the same time as our supposed control over nature has increased, however, we have also realized that we could be operating on borrowed time. Bacteria and parasites are returning with the weapon of resistance polished, we continue to be subject to weather and wind, and, although our population is not only increasing in size but in material demands, we lack an endless supply of resources such as food and water. As nature in untouched quantity decreases, and the human population steadily increases, it becomes ever clearer that in no way have we achieved independence from nature. Should we therefore turn to life itself to find the solution to the eternal battle against pest and plague, and the growing shortage of food, medicine and materials? Our culture is obviously built on taking other species into our service, but perhaps the ultimate solution to many of our challenges lies in the bacterial realm?

On 26 June 2000 'one of mankind's greatest discoveries' was announced: the human genome (our total genetic material) had been mapped. Former president Bill Clinton expressed himself at the time as one would expect in a country that still views Darwin with scepticism: 'Today we are learning the language in which God created life.' The press release from the White House further confirmed that 'Today's announcement represents the starting point for a new era of genetic medicine', while accurately stating that 'the sequence represents only the first step in the full decoding of the genome, because most of the individual genes and their specific functions must still be deciphered and understood.'

A substantial number of people at a substantial number of institutions were behind the discovery, but one person in particular stood out: Craig Venter.[36] The exotic and eccentric, but also brilliant, researcher was crucial to the fact that it was Bill Clinton who came to announce the news in 2000, rather than, for example, George W. Bush in 2005. Venter was the first person to have his own genome fully sequenced and to interpret the genetic message as far as it would allow. He also founded the new field of metagenomics, that is, large-scale studies of the genetic information contained in everything from ecosystems to intestinal systems. When on a cruise in the Sargasso Sea, Venter took water samples and discovered when he analysed them that there was an as yet unexplored world of genetic information out there.

In this context, Synthia proves central as a stepping stone into what some envision as a new world of synthetic biology.[37] Ten years after Clinton's press release came another sensational release, this time from Venter's own research organization, the J. Craig Venter Institute: the first synthetic organism had been created, and this time God was clearly not behind it. Synthia was a bacterium built from scratch using familiar genetic building blocks to create a 1.08 million base pair genome: the largest defined chemical molecule ever created. As Venter himself put it: 'This is probably the first living creature on this planet whose parent is a computer.' And yet, what does all this have to do with carbon?

Quite a bit, actually. Synthia established a kind of bridge between the artificial world and the natural world, and, in terms of life itself, between the living and non-living. It took fifteen years to construct Synthia's genome, a feat that overshadows even the accomplishments of polymer chemistry. It is, however, not Synthia, but rather *Synthias*, all the modified and tailored variants of Synthia, that have proven so relevant. Bacteria can construct a number of different polymers: they can create bioplastics, enzymes and proteins, antibiotics and pigments, to name just a few. It has been many years, for example, since bacterial cells revolutionized the insulin supply when the genes governing human insulin production were inserted into bacteria that read the genetic instructions and immediately began producing insulin. An impressive feat, for sure, but only the beginning.

By adjusting or building on bacterial genetic codes, bacteria themselves can be transformed into biomanufacturers with a formidable production potential. Under favourable conditions, bacteria can split multiple times during an hour and, barring obstacles, will achieve a weight equal to the earth in the course of a few days. One of Venter's great visions is to use bacteria largely to solve the difficulties that global energy production creates for the carbon cycle: for example, by mass-producing bacterial biofuels; by using blue-green bacteria to split H from $H_2O$; and, his ultimate dream, to create a biosynthetic photosynthesis that overshadows every solar panel in existence. If we could create the range of carbon products we require using such unlimited resources as $CO_2$, $CH_4$, water and light as the basis for bacterial production, we could, at the very least, buy ourselves more time.

Synthia's successors have already made the leap from science fiction to industry – especially the oil industry. ExxonMobil has recently invested $600 million in a project to design blue-green algae that can capture $CO_2$ and sunlight for use as bioenergy. Chevron, for its part, invested $25 million in a project to reprogramme the bacteria *Escherichia coli* to break down biomasses for use as biofuels. Such synthetic biology, for its part, largely falls within the field of biorefinery, which again is part of a quickly growing bioeconomy where carbon chemistry, as previously mentioned, unites the concepts of what is natural and what is synthetic. The problem thus far is that Synthia, though partially synthetic, is still a life form, not a silicon chip or something else. Life is unstable and requires nourishment and care. My personal sense is that many of Venter's visions may become realities, but not quickly enough or to the extent that we can simply lean back and wait for them to save us. Unfortunately, replicating the ingenuity plants show in transforming sunlight to an edible form of energy is no trivial task. Even if the entire biochemical factory fits inside microscopic chloroplasts, the process is almost as ancient as life itself and is so complex that we still can only partially understand it, much less copy it.

## TO BUILD AND TO BURN: CARBON IN THE LIFE EQUATION

The atmosphere of Mars is 96 per cent $CO_2$, 1.9 per cent $N_2$ and 0.145 per cent $O_2$. Venus falls into the same category with 96.4 per cent $CO_2$, 3.5 per cent $N_2$ and a barely measurable quantity of $O_2$. Earth's atmosphere, by contrast, stands out with its 0.02 per cent $CO_2$, 78 per cent $N_2$ and 21 per cent $O_2$.[38] Against this backdrop, our focus on carbon might seem strange and our concern for $CO_2$ rather exaggerated. Yet $CO_2$ has a property that $N_2$ and $O_2$ lack, namely, it absorbs long-wave radiation from the earth's surface, therefore making it a greenhouse gas. And for that we should be grateful. Without the greenhouse effect of $CO_2$ (not to mention water vapour, methane and nitrous oxide), earth's average temperature would be –20°C instead of the present ideal state of +15°C. In general, the planet's gas composition is ideal for life: some will even suggest almost miraculously ideal. The reason for this fact, however, is largely life itself, the mutual interplay of life and climate. The atmosphere, that is, has not always been perfect for our needs and today's optimal atmosphere is thermodynamically unstable. It would look very different in the absence of life, just as it did before life gained a solid footing. On a fundamental level, life and climate can be considered as a balance between building and burning, as well as between heat and cold. The relationship, furthermore, between the partners of carbon and oxygen is central for the global thermostat that ensures us such favourable living conditions. Whether or not another planet out there exists with higher life is not only uncertain but improbable, even if statistics in the form of the number of planets seem to favour it. If such a planet does exist, an analysis of its gas composition will probably reveal the presence of life.

In its infancy the earth's atmosphere was chiefly dominated by hydrogen, and probably also water vapour, methane and ammonia (which did not, in this case, have a biological origin). This phase was replaced by a period with significantly higher concentrations of $CO_2$, probably due to volcanic activity, and it was during that period, around 3.5 billion years ago, that life began. Solar emissions back then were 30 per cent lower than they are today, but the high $CO_2$ concentrations apparently compensated for that, enabling temperatures to rise enough to form the basis for life. It was also crucial that the earth's temperature remained high enough to prevent the oceans from freezing. The open ocean absorbed most of the atmospheric 'surplus' of $CO_2$ and precipitated it as calcium carbonate. Although today this chemical carbon pump is still at work regulating the atmosphere's $CO_2$ concentration, back then it must have had an entirely different order of magnitude. The $CO_2$ present in the water also had another effect, nourishing what would come to represent the beginning of the third great change in the earth's atmosphere: photosynthesis.

Carbon and oxygen, of course, are both central to photosynthesis, life's most important reaction, not to mention hydrogen, which is a fundamental element in every sense. These three elements form endless relationships and reactions, but the most important one looks like this:

$$6CO_2 + 12H_2O + \text{energy} \rightarrow C_6H_{12}O_6 + 6O_2 + 6H_2O$$

Many of us have memorized this equation one or more times at school. If evolution had not produced chloroplasts with the ability to harvest solar energy from the sun with its antenna pigments, kneading inorganic carbon and water together to form organic carbon with oxygen as a waste product, any non-plant species on the planet would have had a rough time of it. This is true not only of those that eat greens, but of those that eat those that eat greens, as well as everything that lives by breaking down everything else after death. All species, in other words, that we call heterotrophic, that is, things that get their nourishment for other things (all forms of animals from amoebas to humans, as well as fungi and a significant number of bacteria). Of course, strictly speaking, without photosynthesis plants would not have existed either. It should also be mentioned that producers, or autotrophs (those species able to produce their own nourishment), also receive something, $CO_2$, which is one of the end products of cellular respiration. It is, perhaps, this symbiotic relationship that especially prompts the theologically minded to wonder at how wisely everything seems to be ordered. Whatever the case, wisely or not, the balance between these two inverted processes is what we depend upon.

The effect of photosynthesis is not just the atmosphere as we now know it, which thereby creates conditions for higher life, but a climate that seems wonderfully conducive to providing all life with good living conditions. In all its complexity, it is indeed almost – but only almost – a wonder how this key reaction emerged perhaps as early as 3.5 billion years ago. Exactly when it emerged we will never know, just as we will never know exactly how. Bacteria do not tend to leave behind a notable fossil record, certainly not when it comes to their inner cellular apparatus, though the exception to this rule is stromatolites. These tubers are formed of blue-green bacteria, which, apparently, conducted an early form of light capture, a requirement in times where UV radiation was still mercilessly baking the earth's surface. Yet it is debatable whether the earliest stromatolites offer a final proof of 'modern photosynthesis'. Interestingly enough, active stromatolites still exist, and my colleague Jim Elser makes frequent trips to some spectacular stromalite occurrences in springs in the Chihuahan Desert in Mexico, a relatively unusual field location for a water researcher. These stromatolites, in any case, clearly conduct photosynthesis, and they probably do represent an unbroken tradition that extends back 3 billion years.

BLUE-GREEN BACTERIA were not the only bacteria group to master the art of harvesting energy from light. A number of bacteria groups retained, and still retain, this property. Two groups that are still around today, and that represent the first bacteria to conduct photosynthesis, albeit in the absence of oxygen, are green sulphur bacteria and purple sulphur bacteria. Some lakes exhibit a sharp stratification between oxygen-free bottom water and oxygenated surface water, and these lakes are home to a pink layer of sulphur bacteria just deep enough to escape oxygen's harmful effects, have access to the hydrogen sulphide below them (these water samples seldom smell good), and also receive sufficient light from above. These bacteria also contain several types of bacterial chlorophyll and, generally speaking, more of the reddish pigments that harvest long-wave light in addition to chlorophyll. They also used, and still use, hydrogen and hydrogen sulphide as electron sources instead of water. The result is that these bacteria produce sulphur or sulphate instead of oxygen, something that did not do much for the oxygen level in the atmosphere, which back then was still approaching zero.[39]

Ultimately, around 700 million years would pass while bacteria of various kinds ruled over a sulphur-reeking, grey, lifeless and oxygen-free earth. Without oxygen there was also no ozone layer that could protect the earth's surface against the merciless ultraviolet radiation. As a result, the first bacteria wisely remained beneath the water, which worked to absorb most of this ultraviolet radiation.[40]

Around 2.7 billion years ago, the silent oxygen revolution began that would completely change the face of the earth. It was an endlessly slow revolution, but once the terrible toxicant, oxygen, began to accumulate, the consequences were dramatic: toxic, that is, to ancient earth's organisms, which were dependent on an oxygen-free atmosphere and which did not have a cellular apparatus equipped with protective pigments and anti-oxidants that could manage oxygen's potentially devastating effect within their cells. Oxygen is not harmless to us either, and even though we have developed advanced mechanisms for coping with free radicals and other such nefarious oxygen products, oxidative damage is one of the reasons that cells age, that cell membranes become less supple, and that the skin wrinkles. Ancient earth's organisms would continue to rule the surface for another billion years or so, but were also steadily relegated to the remaining pockets with oxygen-free conditions, such as deep lakes and seas, deep within the soil and swamps, inside intestinal tracts and other places where an anaerobic environment persists.

Today oxygen accounts for 21 per cent of the earth's atmosphere, which is 'just perfect'. Much more oxygen and the danger of forest fires would substantially increase: at 35 per cent atmospheric oxygen, forest growth could hardly keep pace with the frequent fires. Simultaneously, our cellular machinery would experience a significant increase in stress due to oxidant damage. Notably lower amounts of oxygen, on the other hand, would result in breathing difficulties. Our respiration depends upon oxygen's concentration in the atmosphere remaining as it now is, something that is not random, but is no miracle either.

## FROM C3 TO C4 – AND THE WORLD'S MOST IMPORTANT PROTEIN

How did nature and evolution develop such a complex and apparently miraculous photosynthesis, and who 'discovered' it?

Even if forms of photosynthesis in plants, algae and blue-green bacteria have many differences, a common characteristic is that intricate, organic molecules are built with help from solar energy via antennae (pigments) that harvest photons.

Given that it took several hundred million years before nature developed photosynthesis as a way of obtaining energy, it is perhaps not so strange that we have not entirely managed to replicate photosynthesis synthetically, even if that is the dream of many. Despite what its simple equation might lead us to believe, the photosynthesis reaction is far from simple.[41] The reaction can be divided into two parts: the light-dependent reaction and the light-independent reaction (the dark reaction). For our purposes, it is the photosynthesis that utilizes $CO_2$ and produces $O_2$ that is more interesting. Here water is oxidized into oxygen and hydrogen ions. Oxygen, as a by-product, is separated from the cell, whereas electrons from the oxidized water molecules are transferred to a reaction centre consisting of chlorophyll and different proteins (photosystem II), where they are further transported to a number of proteins (the electron transport chain), before they are again conveyed to a chlorophyll-rich reaction centre (photosystem I). Through a series of reactions involving various different enzymes, we finally end up with the energy-bearing molecule ATP. ATP is key in dark reactions where the enzyme rubisco directs the pivotal transformation of $CO_2$ into, for example, sugar or starch. If you think all this sounds complicated, be aware that it is an almost shameless oversimplification.

An old question, which is as difficult to answer as it is easy to ask, is: what is life? Where do we draw the boundary between non-life and life? At this point we can content ourselves with pointing out the fundamental transformation that happens in photosynthesis between inorganic and organic carbon. This is where rubisco comes into play, and it is a protein that definitely warrants a few sentences. Rubisco (Ribulose-1,5-bisphosphate carboxylase/oxygenase) is not only the world's most common protein, it is undoubtedly the world's most important. It is primarily rubisco that performs the magic trick wherein $CO_2$ undergoes the decisive transformation from atmosphere to biosphere, from $CO_2$ to organic C in the form of sugar. If we had access to a form of synthetic rubisco that could perform this trick for us, that could extract C from $O_2$'s intense embrace and could connect it to all our combustion processes, *then* we would be talking.

At the same time, rubisco demonstrates that even evolution has its limits and does not always reach the perfect or optimal solutions to meet life's challenges. As such, it can be parenthetically remarked that rubisco, for all its apparent wisdom, is a good argument against intelligent design. Like so much else, rubisco is a child of its time, and it was actually developed by blue-green bacteria some 3 billion years ago. Rubisco remained relatively unchanged through plants' evolutionary cycle, and functioned exceptionally well as long as there was ample $CO_2$ (and little oxygen). The problem with rubisco, namely, is that it exhibits an affinity not only for $CO_2$, but for oxygen, and when the relationship between these two molecules shifts in favour of more atmospheric $O_2$, rubisco proves less effective. Rubisco was developed in an atmosphere that contained around a hundred times more $CO_2$ than today, not to mention hardly any oxygen. Since the entire photosynthesis reaction, in which rubisco plays a central role, contains a number of key-fits-the-lock adjustments, it has not been possible to return to the drawing board and develop a new enzyme that is better suited to today's reality without redesigning large portions of this complex process.

There is, nonetheless, a way around this problem and the plants that have discovered it are called C4 plants.[42] The oldest plants on earth are C3 plants, so-called because they first transform $CO_2$ into three carbon compounds. These plants literally blossomed 250 million years ago, in the age of the reptiles, even though the first vascular plants actually appeared 200 million years earlier. These C3 plants crave the highest $CO_2$ concentrations possible, and even though they have long existed at the edge of famine, when it comes to $CO_2$, at least, they still dominate the world's plant populations. C4 plants are younger, and as you have probably guessed, have the ability to transform $CO_2$ directly to molecules with four C atoms. In addition, C4 plants can increase $CO_2$ concentrations inside cells where rubisco comes into its own and transforms $CO_2$ via organic acids into sugar. This process exploits carbon much more effectively, and it is not without reason that our most productive food crops, such as corn, sugarcane, millet, sorghum and most grasses (not to mention stubborn weeds), are C4 plants.

C4 plants emerged during a series of ice ages around 4 million years ago when $CO_2$ levels sank to a critical (for C3 plants) 200 ppm and the climate was cold. Extensive areas on earth became dominated by C4 grasses, which have since proven crucial actors in the earth's land-based ecosystems. Only 4 per cent of earth's plants are C4, but they represent 21 per cent of the photosynthetic activity on land – and that even though only one C4 plant can be classified as a tree. The chronic carbon famine that C3 plants experience is also the reason why these plants grow more quickly with higher levels of $CO_2$, something that might lead one to believe that increasing amounts of atmospheric $CO_2$ would have positive effects, for example, through increased C3 plant production. And this does occur. Most plants enjoy significantly more growth when more $CO_2$ enters the picture, and that is probably one of the reasons why the atmospheric $CO_2$ content is not increasing at an even greater pace. It remains to be seen which of these plant groups stands to gain the most with a changing climate. Whereas increasing amounts of $CO_2$ are an advantage to C3 plants, increasing temperature and greater drought should benefit C4 plants, although these also have limits to their endurance when it comes to temperature and drought. As a result, for large portions of the globe a changing climate will prove negative for both groups.

Plant growth, furthermore, is not just $CO_2$-dependent, so even if we add more $CO_2$ (and nitrogen) plant growth will also be curbed by water shortage or by phosphate deficiency, a subject to which we shall return. Plants also have a tendency to reduce their nutrient content with the single-sided addition of $CO_2$. In his extremely influential work *An Essay on the Principle of Population* (1798), Thomas Robert Malthus underscored the fact that famine catastrophes were a real danger because population growth exceeded the growth in food production.[43] What Malthus could not have foreseen was the green revolution, the dramatic increase in the world's food production that came about as a result of irrigation and fertilizers. Because of this Malthus was branded, rather undeservedly, as the world's first environmental pessimist. Today there is a hope of renewing and extending the green revolution through different forms of genetically modified plants: one of the most fiercely promoted dreams is to equip C3 plants, for example, rice and wheat, with C4 plants' turbophotosynthesis. This idea straddles the boundary between science and fiction, a boundary, however, that is by no means sharply delineated. If this dream proves successful, then we are truly talking about a new green revolution, although no tree grows in the sky, quite literally in this case. Plants require other things to grow, and water and phosphorus are essential. Water is already a minimum factor in the earth's agriculture, and phosphorus might become one in the uncomfortably near future.

WE HAVE also not answered the question as to how photosynthesis first emerged. How did all the gears in this complex machinery come to fit each other? Answering this question is like answering the question of how life began. No one knows the exact details, but like most complex phenomena, its emergence was modular. Rubisco also shows us, as stated above, that although the process is impressive, the optimal solutions are not always the ones selected. Parts of the reaction probably had

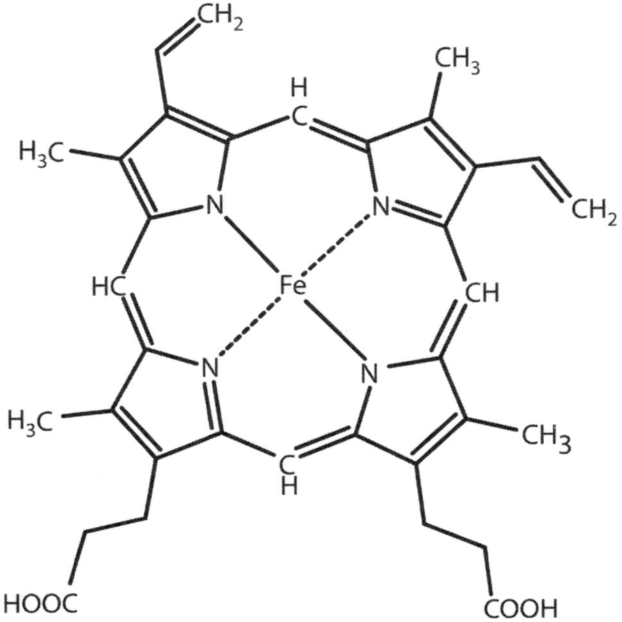

The heme molecule and the chlorophyll molecule are evolutionarily tailored in the same manner, perhaps with porphyrines as a point of departure for the evolution of the more complex molecules.

other functions. Take chlorophyll molecules, which in all their complexity have a long series of relatives that are all porphyrins. When it comes to porphyrins, we still find the ever-present carbon hexagon, the basal building block, although here the nitrogen atom is centrally placed. Basically stated, porphin and its variants, when connected together into larger molecules, form the porphyrins.

Our own heme molecule, the main component in our blood, is in this same family. The chlorophyll molecule, on the other hand, is structured around magnesium and has the chemical formula $C_{55}H_{72}O_5N_4Mg$, whereas blood is structured around iron, for example, heme A: $C_{49}H_{56}O_6N_4Fe$. As we can see, this compound is strikingly reminiscent of chlorophyll. It is typical for nature to vary themes according to the same basic principles. Characteristic for all porphyrins is that they are experts at capturing light, which is why they are often strongly coloured (*porphyros*, 'purple'), and it is a short jump from there to molecules that exploit light's energy, that is, the antenna pigments that capture electrons.

What we do know is that at some point in history one of these blue-green bacteria formed an eternal partnership with another primitive cell.[44] This primitive cell 'consumed' the blue-green bacteria, and it turned out that the bacteria was more useful as a partner than as a food item. Together these two achieved a unity capable of translating the sun's energy to chemical energy in the form of sugar or other organic molecules. The foundation for the plant kingdom was thereby laid – and the rest is history.

So much for the evolutionary *invention*, but what about the *discovery*? Here we might start with the ancient Greeks. Aristotle is credited with the insight that plants, too, require nourishment, though he was hardly the first to realize that. People had already been cultivating food crops for 8,000 years, and had undoubtedly noticed that the application of water and manure yielded better crops. On the other hand, Aristotle elegantly formulated the basic insight that plants did not need animals, but animals needed plants. Strictly speaking, all animals, ourselves included, more or less freeload off the plant kingdom, even if we do give back $CO_2$, since there are other, more important sources of $CO_2$ than animals (bacteria, for example, or volcanoes).

In the mid-seventeenth century it was speculated that plants might take in 'something' through their leaves, and it was obvious that plants did not thrive in the dark. As such, there must be 'something' to light. However, as we have seen, it was not until the latter part of the eighteenth century that Joseph Priestley grasped the basic principle: into plants goes $CO_2$ and out comes $O_2$. Light plays a role somewhere in there, but where? The subsequent history regarding a full understanding of photosynthesis has been long and laborious, and, strictly speaking, we have not entirely achieved the goal. Nonetheless, it is thanks to atomic physics that photosynthesis's basic principles were recognized.

## ATOMIC PHYSICS'S SOLUTION TO PHOTOSYNTHESIS

The history of science is full of striking examples of the adage that two heads are better than one, that the communal exchange and analysis of thought can take us further than the lonely genius sitting in his lab, even if science history also contains successful examples of the latter. Late in the 1930s Samuel Ruben and Martin Kamen arrived in Berkeley, California, where they quickly united in the mutual, ambitious goal of solving photosynthesis's mysteries.[45] In order to understand how carbon made its way from $CO_2$ into other forms, they reasoned, it would be useful to mark the carbon atoms in some way, but how to mark an atom? The most obvious method was to use isotopes – varieties other than the standard 12C, that is – but there was not exactly a wealth of options. The only radioactive form of carbon was 11C, but the problem was that 11C has a half-life of only 21 minutes. That meant that everything, both experiment and analysis, had to be done with lightning speed before the radioactivity faded, and Ruben and Kamen recognized that they needed an isotope with a longer lifespan.

Most people would have resigned themselves to this eventuality – after all, new isotopes don't exactly leap off the shelf – but Ruben and Kamen took matters into their own hands. Berkeley provided them with access to a radiation lab equipped with a cyclotron, and after long experimental stings with only weak results, Kamen returned to the lab on 19 February 1940 and for the next 120 hours bombarded graphite with electrons in the cyclotron. Finally, late in the evening, he collected the pieces of bombarded graphite, placed them on Ruben's desk for later analysis, and went home to grab some sleep. The next morning he was awakened by an excited Ruben, who had tracked the occurrence of a new, radioactive form of carbon. This would turn out to be 14C, an isotope with six protons and eight neutrons, and, most importantly of all, a prodigious half-life of 5,730 years. In principle there was now a way to solve the mysteries of photosynthesis, but the Second World War intervened, temporarily halting all further experiments. Ruben was recruited for the war effort, and began studying the effects of the poisonous gas phosgene on lung proteins. Phosgene, by the way, has nothing to do with either phosphorus or genes, but is actually carbonyl chloride, one of the deadly alliances formed between carbon and chlorine, also joined here with oxygen ($COCl_2$). Ruben did not develop this gas, whose terrifying effects had been thoroughly documented in the First World War. What interested Ruben was the gas's physiological effects. He used the 11C isotope here to document how the gas bound itself to lung proteins, and on 27 September 1942 became an involuntary testament to phosgene's deadly effects when he was exposed to a high dose and died the following day, at not even thirty years old. With that, Ruben joined the ranks of scientists who have died at their posts, a victim of his chemistry, but also of the war. Kamen was also a victim of the war: he was banned from the radiation lab, which was now being used to develop an atom bomb.

The new age brought with it reactors capable of producing large quantities of 14C, and it was subsequently the trio of Melvin Calvin, Andrew Benson and James Bassham who fulfilled the dream of uncovering photosynthesis's secrets. It was now possible to follow $CO_2$ molecules that had been marked with $14CO_2$ through the entire photosynthesis cycle, and to analyse the path the radioactive carbon took through different compounds before finally ending up as radioactive sugar.

It was, moreover, this same principle and this same radioactive carbon that I utilized in my efforts to understand the carbon cycle in that forest lake near Oslo. Among other things, I added radioactive carbon (in tiny, harmless doses) to small bottles that were then distributed in the water at various depths. After several hours, I was able to filter out the algae that had become trapped at the varying depths, with varying light exposure, and to analyse their radioactive 14C content. Since I knew beforehand what the proportion of 14C was in relation to 12C (a very small fraction), I could calculate exactly how much $CO_2$ the algae in each sample had absorbed: a tiny amount, to be sure, but a precise snippet of the global carbon cycle.

Calvin's cycle, also known as the Calvin-Bensen cycle or dark reaction,[46] can briefly be described as the process whereby $CO_2$ binds to an unstable 6-carbon compound with help from the enzyme rubisco. After the cycle has run its course six times and has accumulated enough energy, the compound stabilizes in the form of glucose ($C_6H_{12}O_6$). Glucose can form the basis for cellulose or starch, or can be transported to the cell's mitochondria, where energy is again released through cellular respiration.

Today Calvin's cycle is a basic component of biology textbooks, and Calvin was awarded the Nobel Prize for Chemistry in 1961. It was a highly deserved honour, even though many thought that Kamen and Benson ought to have shared the podium. It was, incidentally, at a White House lunch with the Kennedys that Calvin and other Nobel Prize laureates were treated to Kennedy's famous remark, 'I think this is the most extraordinary collection of talent, of human knowledge, that has ever been gathered together at the White House, with the possible exception of when Thomas Jefferson dined alone.'

In any event, the main characteristics of photosynthesis were thereby solved, and new details have since emerged, but there are still aspects of photosynthesis that are not entirely understood. What we do know is that all plants build organic material from $CO_2$ and water with the help of solar energy and rubisco in a complex, microscopic factory. Thanks to this process, the rest of us, we who have not mastered the art of photosynthesis, can exist. Perhaps it is not entirely correct to say that we freeload off plants, because thanks to a number of combustion processes we do return large quantities of $CO_2$ to them. Of course, plants carry out both sides of the process, and can both build and burn. Throughout time the globe's climate has been and will be largely regulated by the balance between building and burning – that is to say, by the relationships into which carbon enters.

## PHOTOSYNTHESIS IN REVERSE

One hot summer's day by the inland forest lake where I run my experiments, I dropped my water sampler and it sank quickly to the bottom. It was close to shore and shallow enough for me to make out the sunlight reflecting off the metal below. It appeared a golden reddish-brown, the colour of the water, because all the dissolved, organic carbon in the water quickly absorbs shortwave light, the ultraviolet, blue and green, for example, whereas yellow and red penetrate deeper. A water sampler is not something you want to lose, and I had just measured the surface temperature to be at 22°C, so there was not much to hold me back. I took off my clothes and dived in. At the bottom it was ice-cold, like sticking your head into a refrigerator, as I grabbed the water sampler and returned to the heat and the light.

A lake is ideal for illustrating a great many things. Take the effect of light. Light equals photons, and it can drive photosynthesis or create heat. Both processes are happening on the lake's surface. Some photons are harvested by the brown, dissolved carbon in the water – or by the water itself – and some by the algae's chloroplasts. This ensures warmth on the surface, though there is little to warm up the depths. Therefore, the water sampler's thermometer showed 10°C when I retrieved it from the bottom, whereas it was a comfortable swimming temperature above. There is also a corresponding transition from the surface, where organic matter is produced, and in the depths, where it is broken down (respirated), for the same reason: there are not enough photons left to drive photosynthesis in the depths. When I measured the lake's oxygen profile, I saw that $O_2$ decreased quickly from full saturation on the surface to an oxygen-free area 10 m down. $CO_2$ demonstrated the opposite progression, from small quantities 5 m down to a strong increase at the bottom, where large concentrations of $CO_2$ also keep company with methane and hydrogen sulphide. This fact demonstrates how life's two most important processes are also mirror images of each other.

Whereas photosynthesis builds complex molecules with the help of solar energy, respiration or cellular respiration breaks down the energy stored in complex molecules in order to reverse the process. In its simplest form, the inverse or mirror image photosynthesis equation reads:

$$C_6H_{12}O_6 + 6O_2 \rightarrow 6H_2O + 6CO_2 + \text{energy}$$

Whereas sunlight or photons provide the energy that drives photosynthesis to produce an organic and combustible (oxidizable) organic molecule, it is the combustion or respiration of that molecule that releases the energy stored in the form of adenosine triphosphate, ATP. Once again, anyone who remembers their school lessons will recall that ATP is created through complex cycles, but here we need only focus on the fact that these processes have the amazing property of balancing one another in a global symbiosis between producers and consumers. It is true that plants also drive respiration and produce $CO_2$, and at night plants will release more $CO_2$ than they absorb, but over time a plant will absorb more $CO_2$ than it releases – otherwise, it would simply fail to grow.

We do not need to rely solely on sugar to fire our cells. We can put any number of things on the hearth to create heat (enough to maintain 37°C) and energy for life's daily tasks, whether that be writing a book on carbon or taking a jog. And that hearth? It is mitochondria, the cells' power station, which acts as a parallel to chloroplasts. In this case it is another complex cycle, the Krebs or citric acid cycle, that creates energy. Our average muscle cells contain between two and three hundred mitochondria. Long before the Cambrian explosion, probably as long ago as 1.5 billion years, cells emerged that now contained a nucleus for their DNA. Up until then all life had consisted of bacteria, which by definition are cells without nuclei. One of these early nucleated cells swallowed a free-living bacterium, but instead of digesting it, a symbiotic relationship occurred, an eternal partnership. This history is akin to the origin of chloroplasts, but it happened earlier (plants also have mitochondria) and probably only once. Because mitochondria and chloroplasts have their own DNA, it is possible to track this change, and in the case of mitochondria the signs point toward a group of bacteria in the SAR 11 family (SAR after the Sargasso Sea, where they were first described), the descendants of which still exist as free-living bacteria.

Primitive mitochondria had an invaluable talent: they could produce energy through the combustion of organic material with the help of oxygen. As a result, they put troublesome oxygen to use, thereby producing energy in a significantly more effective way than the old, oxygen-free metabolism could. This was an extremely useful partnership, therefore, and one on which all higher life subsequently built. Plants established themselves by having both an apparatus to transform $CO_2$ and sunlight into stored energy, that is, chloroplasts, and a combustion apparatus to exploit that same energy (mitochondria). Animals, fungi and all other heterotrophic life are equipped with mitochondria, and acquire energy by eating or freeloading off chlorophyll-equipped life. This vital balance between plant oxygen production and our $CO_2$ production is, in reality, the balance between energy production among two ancient types of bacteria: those that gave rise to the chloroplasts and those that gave rise to the mitochondria.

Fire is not a picky eater; it does not necessarily require dry pinewood, but will accept most organic material once it gets going. If we look at life as a whole, there is almost no limit to what can be combusted. Bacteria and fungi in particular have an appetite for most things. Different groups and species, of course, tend to each have their own speciality. Everything that gets broken down, that decomposes, does so with help from bacteria and fungi, often working in tandem. As such, we can count on many toxins, plastics and oil to disappear eventually. Microorganisms also have the great potential to adjust their metabolisms as needed by making swift, evolutionary adaptations. As a result, the oil that spilled into the Gulf of Mexico in 2010, despite all the damage it did to flora and fauna, eventually disappeared owing to the numerous bacteria that can use crude oil as an energy source with $CO_2$ as the final product.

We ourselves play on a more limited register, using proteins, carbohydrates and fat as our fuel. Carbohydrates are the most effective fuel, with glucose, $C_6H_{12}O_6$, at the top of the list. When there is a shortage of carbohydrates we turn (on a mitochondrial level) to fats, and last of all to proteins. No matter whether we consume broccoli or bacon, the food has two purposes: to build and to burn. The difference from plants is simply that we must extract fuel from elsewhere, whereas plants manufacture their own fuel. The fuel principle is nonetheless the same: solar energy being converted to stored energy through photosynthesis. In the case of broccoli, it is probably a year's worth of stored solar energy, but when you drive to the supermarket to buy broccoli, you are being propelled by solar energy that has remained locked away for several hundred million years.

## GOOD NEIGHBOURS

One of the most striking things we learned from our experiments at the lake, where we followed carbon's wanderings in and out of the ecosystem's many actors, was all the different shapes and forms carbon took. In the first place, there was the great quantity of dissolved, dead carbon in the form of humus (or humic compounds) floating in the water. Around 90 per cent of all the carbon in a typical forest lake is just such humus, broken-down plant remains from land, easily recognizable by its brown colour. If we collect all the biomass present in plankton, ground-dwelling organisms and fish, we are left with only 10 per cent of all carbon in the water. Within the organisms themselves, of course, there is still an overwhelming amount of carbon in comparison to life's other components. We also discovered higher concentrations of $CO_2$ and methane ($CH_4$) in the water, something that showed that lakes can be significant sources of these two greenhouse gases. In short, carbon figured as the particularly dominant element in every context.

When we examined the relationship between carbon and other elements, we also found some conspicuous patterns. In the first place, the relationship among elements varied between species, and also between plants and animals. Plants contained a significantly higher quantity of carbon (C) and lower quantities of nitrogen (N) and phosphorus (P). Within a single species, and especially among animals, in contrast, there was very little variation among elements and that put us on the track of an important mechanism: plant-eaters often ingest far more C than they can use, simply because there is so much more of it than the other elements critical to life, such as P or N. These latter elements, as well as others that occur in even smaller quantities, can be limiting for growth among plant eaters. In order to maintain a stable relationship between C, N and P, animals must discard the excess carbon they ingest. Some can be stored as fat, but animals can also rid themselves of this carbon excess by increasing their respiration.

Humans, we well know, cannot live off bread alone and, as we discovered, plants cannot live off carbon alone. Carbon is located at the top right of the periodic table. Its closest neighbour, with an atomic number of 7, is nitrogen, and after that oxygen. In the next row, we find silicon, phosphorus and sulphur. These good neighbours particularly function as life's building blocks: C, N and P especially. Oxygen (O) is also found in this exclusive neighbourhood and largely proves a building block in most compounds in the form of O, however, and not as $O_2$. Oxygen comprises about 65 per cent of the total weight in humans and other animals (slightly less in plants), and even after water is taken out of the equation, O amounts to around 35 per cent of that weight, not too far behind C. Sulphur figures in a wealth of enzymes, hormones and biochemical reactions. If, in this context, we limit our remarks about O and S to these few lines, it is not because these elements are unimportant, but because they are not life-limiting in the same way as the others.

The stable relationship between the body's essential components is not random, but reflects how these elements are used. C is used for almost everything. Carbohydrates are composed of C, H and O, but not of N and P. As such, we do not need N and P to build carbohydrates, but we also do not get any N and P by eating them. Fat mainly consists of C, H and O, but can also yield a small amount of N and P. Proteins, like everything else, have C, H and O, but also a good quantity of N, whereas nucleic acids like DNA and RNA have C, H and O, but also quite a bit of N and P.

As a result, C is ubiquitous, just like H and O. N is a key element in proteins, and typically accounts for 10 per cent of the dry weight in humans and other animals. The need for C and N is therefore obvious, but what about phosphorus (P)? This element comprises just over 1 per cent of our dry weight, but is still

| | | | | | |
|---|---|---|---|---|---|
| | | | | | 2 He Helium |
| 5 B Boron | 6 C Carbon | 7 N Nitrogen | 8 O Oxygen | 9 F Fluorine | 10 Ne Neon |
| | 14 Si Silicon | 15 P Phosphorus | 16 S Sulphur | 17 Ci Chlorine | 18 Ar Argon |
| | | 33 As Arsenic | 34 Se Selenium | 35 Br Bromine | 36 Kr Krypton |

In the upper-right corner of the periodic table, carbon is in good company with the other key elements of life, such as nitrogen, oxygen, phosphorus, silicon and sulphur.

a key element. Without phosphorus, there would be no DNA, no RNA capable of translating DNA's messages into protein, no ATP (the central energy-bearer in cellular respiration), and no vital phospholipids in the cell's membrane.

In short, life as we know it would not be possible without phosphorus, and in the search for extraterrestrial life, it is not enough to search for water or liveable temperatures. C, N and P must also be present. Paradoxically, phosporus's importance is due to its scarcity. Here, like everywhere, it is a question of supply and demand, and even if we require only modest amounts of phosphorus, it can become critical if the supply decreases even more.

It should be mentioned here that others have reflected on this fact; it is difficult to make *entirely* new discoveries in our day and age – varieties of gunpowder have often been invented before. Our old friend Jöns Jacob Berzelius was the first to use the concept stoichiometry, from the Greek *stoikheion* ('element') and *meter* ('measure'), to refer to the proportional relationship among elements in materials and chemical reactions.[47] For example, it is not at all random that it takes twelve oxygen atoms (six $O_2$) to burn one glucose molecule, and that this reaction forms six water molecules and six $CO_2$ molecules. It is the minimum requirement and the minimum product when transforming $C_6H_{12}O_6$. The fact that the number six again figures in this equation is due to glucose's stoichiometry, which is $C_6H_{12}O_6$.

We have also mentioned Justus von Liebig. He was the person who inspired the discoverer of the benzene ring, August Kekulé, to switch from architecture to chemistry, perhaps because both fields focused on structure and connections. Liebig also taught Friedrich Konrad Beilstein, the man who started the endless task of classifying all organic chemical compounds. And I must, without implying any type of comparison, say that Liebig – or rather, Liebig's insights – have proven a crucial inspiration for my own research into biological stoichiometry.

Liebig was one of those people who wanted his childhood room filled with 'my first chemistry set', firecrackers and other things with sizzle. His excitement for the market clowns of his time, with their fireworks and mercury fulminate demonstrations, gave him a lifelong love of chemistry. He completed his doctorate in 1823, just after turning twenty, and the following year, on the recommendation of none other than Alexander von Humboldt to the Grand Duke of Hessen, Liebig was given a professorship in chemistry and pharmacology at the University of Giessen. His credits range from baking powder, baby food and beef extract to chloroform, free radicals and, not least, plant fertilizer. It is the last field that is relevant here. Since our theme is carbon, however, we might also add the *kaliapparat* to Liebig's list of accomplishments. In 1831 Liebig invented a device comprising five round glass bulbs connected together. In order to determine the carbon quantity in organic compounds, Liebig burned the compounds and circulated $CO_2$ gas through the reaction chambers, until finally he ended up with potassium carbonate. He used this final product to calculate the amount of carbon in the original compounds before they were burned. This extremely clever device was also the world's first instrument for $CO_2$ capture.

Liebig, moreover, understood that plants did not just require $CO_2$; they also needed nitrogen and a number of other compounds to achieve optimal growth. Furthermore, one element at a time would also prove limiting. If the soil contained too little phosphorus, it did not help to pour on other elements and compounds such as nitrogen or $CO_2$. Liebig's Law of the Minimum can be represented as an old-fashioned water barrel made of vertical staves. The quantity of water in the barrel will be determined by the height of the shortest stave, often phosphorus, and the barrel will not hold more water, even if the other staves are lengthened. The barrel metaphor reflects the fact that plants also crave elements in a certain proportion in order to optimize their growth.

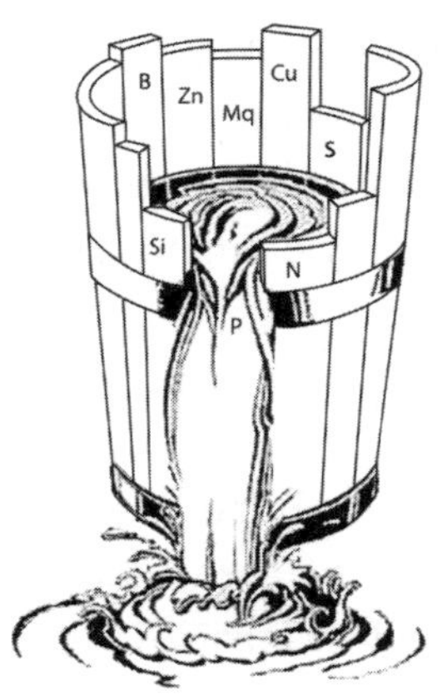


Liebigs's bucket illustrating the Law of the Minimum. The productivity of an organism or an ecosystem is represented by the water level in the bucket and always limited by the shortest stave (representing a specific element, often phosphorus).

We are no different. The stoichiometric equation for a human being, which takes into account the body's water content, will look like this for the most important elements:

$$H_{375\ 000\ 000}\ O_{132\ 000\ 000}\ C_{85\ 700\ 000}\ N_{6\ 430\ 000}\ Ca_{1\ 500\ 000}\ P_{1\ 020\ 000}\ S_{206\ 000}\ Na_{183\ 000}\ K_{177\ 000}\ Cl_{127\ 000}\ Mg_{40\ 000}\ Si_{38\ 600}\ Fe_{2\ 680}\ Zn_{2\ 110}\ Cu_{76}\ I_{14}\ Mn_{13}\ F_{13}\ Cr_{7}\ Se_{4}\ Mo_{3}\ Co_{1}$$ [48]

CHEMICALLY SIGNIFICANT here is the fact that human beings are diverse. We contain a significant portion of the periodic table and deficits in any one of these elements, for example, iron (Fe) or iodine (I), will lead to problems, in this case anaemia and goitre, respectively. Otherwise, this formula tells us that there are 375 million H atoms per cobalt atom, and that life is aquatic, something demonstrated by the dominance of H and O. If O and C dominate by weight, it is because H has the modest atomic mass of 1, whereas O and C are 16 and 12 respectively. We can depart from this stoichiometry, but not by much. Building a human being requires maintaining a certain proportion between the central neighbouring elements C, N and P. If we only ingest carbohydrates, we will not be short on energy (on a purely chemical level, those who live on this diet will hardly feel energetic), but will lack N, P and other important building blocks. The result is that we will have to discard a mass of surplus C, or even worse, we will store it as fat.

As we will eventually see, this stoichiometry also has great significance for the carbon cycle. In 1934 the oceanographer Alfred Redfield characterized the close neighbourly relationship that C, N and P have in the ocean with the formula $C_{106}N_{16}P_{1}$. What surprised Redfield was that all the measurements he took in the ocean seemed to have this composition, something that ultimately became the iconic Redfield ratio. As it turned out, the empirical basis for Redfield's ratio was actually rather flimsy. Even more surprising, however, was that after several decades passed and the number of measurements soared, Redfield's ratio was confirmed – with a certain amount of deviance, true, but nonetheless the ratio was almost magically stable – not only in the water but in particles.

We are still exploring just how this fact occurs, because there is obviously a kind of mutual regulation among these three neighbours that is also reflected in the ocean's algae. One study my colleagues and I did a few years back, based largely on available data, gave a somewhat revised elementary compound: $C_{166}N_{20}P_1$.[49] Exactly how much the increased carbon content already signals a more $CO_2$-rich atmosphere is uncertain, but it also signifies that one P atom can bind 166 C atoms.

More nitrogen and more phosphorus, therefore, indicates more carbon (and $CO_2$) trapped in forests and algae. In many open ocean areas, iron is actually limiting, which prompted the researcher John Martin to announce with uncharacteristic emphasis to a group of researchers-to-be, 'Give me a half tanker of iron, and I will give you an ice age.' Implicitly, by adding large quantities of iron to areas with iron limitation, algae will bind so much $CO_2$ that warming will be reversed. This is not entirely true. In July 2012 100 tons of iron sulphate were scattered from a fishing boat off Canada's west coast.[50] The goal was to increase the algae content, thereby creating a larger fishing ground and simultaneously capturing $CO_2$. The results were not unambiguous. A localized algae bloom did occur, but most of the iron ended up diluted and sedimented. Just like other proposed quick fixes – whether that be space mirrors launched into satellite orbit and used to reflect the sun, supertankers loaded with iron, $CO_2$ stored in rocks, or a number of other more or less fantastic ideas – ocean fertilization with nitrogen, phosphorus and iron will not solve our climate problems, at least in the space of time we have left.

Apropos of quick fixes: wood contains about 50 per cent carbon by dry weight, but only 3–4 per cent nitrogen and only 0.1 per cent phosphorus. That means that an increased supply of

N and P here will have an even greater impact in terms of $CO_2$ uptake (iron is not limiting on land), which has caused some to campaign for forest fertilization to achieve a double benefit: more trees and less $CO_2$. Now it should be noted that there has already been a massive and unintended fertilization of sea and land such that we have fundamentally affected the nitrogen and phosphorus cycles and have set more of these elements in circulation. For example, large parts of Europe currently receive up to ten times more nitrogen from precipitation than was the case in pre-industrial background values, and the supply of phosphorus from airborne particles has also increased significantly. This phenomenon is not limited to Europe: a global fertilization experiment is occurring, and paradoxically this will contribute to counteracting our influence on the carbon cycle. It does at least help with side effects in terms of changes in the ecosystem and shifts in species composition. One of the by-products of the nitrogen cycle, however, is nitrous oxide ($N_2O$), a much more potent greenhouse gas than $CO_2$. Increased nitrogen fallout as a result of human activities has increased the production not only of nitrous oxide, but of methane from a number of ecosystems, and therefore forest fertilization with nitrogen can turn out to have a boomerang effect, resulting in less $CO_2$ but more $N_2O$. This has no net gain for the climate, and, as such, is no shortcut either.

There are many lessons to draw from this idea. Foremost, however, it demonstrates carbon's intimate interaction with its neighbours, as well as the significant consequences this interaction poses for life and climate, not least due to the fact that the carbon cycle is connected to other great cycles. It also functions as a warm-up for a deeper dive into the carbon cycle, into the roles played by forests and oceans, and into what is perhaps the most important of all: the fact that not only does climate impact life, but life impacts climate. This becomes especially obvious when we look at another of carbon's alliances, which is used in a little bit of everything, but which is mainly important because it is the wild card in global climate development.

## C AND 4H: A HOT PARTNERSHIP

Back when I was spending days and nights at the forest lake, it was in order to understand the lake's internal carbon cycle via food chains. Along the way I saw that the carbon circulating in the lake largely originated in the surrounding swamps and forests, from the $CO_2$ that the firs converted to needles, branches, trunk, roots and soil through sugar, starch, lignin and cellulose before some of it eventually ended in the lake as the brown, humic compounds. What was truly surprising was that the lake (like most lakes, as it turned out) first and foremost released $CO_2$ into the atmosphere. Some of the carbon, which at one time was trapped in fir trees, swamps and soil, ended in the water where an army of hungry bacteria threw themselves on it, and the end product of this feast was $CO_2$ gassed back into the atmosphere. This comprised a single patch of the great cycle that, strictly speaking, is a patchwork of the many smaller cycles. The essence is nonetheless the same: some things build (photosynthesis) and others burn (respiration). There was one thing, however, that did not entirely fit the equation, and that was that, quite simply, too much $CO_2$ was being released back into the atmosphere. There must be other sources of $CO_2$, and I returned to the lake to discover them.

Few places are as still as a small forest lake on a sun-filled August afternoon. There is not a breath of wind; the water's surface reflects the sky and appears bright blue, though beneath the surface the water is brown, coloured by all the dissolved organic carbon from trees, swamps and soil. Dragonflies patrol the reeds and a handful of cloudberries grow along the water's edge. When I go to pluck them, my steps cause a bubbling to occur in the swaying swamp along the water's edge. It is the content of these bubbles that I am going to investigate. I take water samples every half-metre down, filling bottles plugged with gas-tight corks. The deepest water samples have the unmistakable odour of sulphide, as one would expect when oxygen is absent. From previous experiments I know that there is a lack of oxygen down there, because I have also taken $O_2$ measurements at the same half-metre intervals many times. In a variety of places I have also sunk an inverted funnel with a water-filled bottle screwed fast on top. When the gas bubbles up from the bottom and into the funnels, the gas displaces the water, and after a time it can be calculated how much gas is released per square metre. I will return here many times to repeat these measurements, including beneath the ice in winter.

When we later analyse the gases in the water, it comes as no great surprise that there are high concentrations of $CO_2$, especially in the depths. We also find $CH_4$, methane. That is not entirely unexpected either – it is the gas that bubbles up from the water's edge, after all – but there is so *much* methane, especially down at the bottom. Yet the methane content quickly decreases toward the surface, and even though this lake, along with most other lakes with oxygen-free sediment, as well as swamps and wetlands, is a significant exporter of methane into the atmosphere, most of the $CH_4$ is converted into $CO_2$ on the way up from the depths. There is a tiny food chain that we can watch here: down in the oxygen-free depths there are methanogenic (methane-producing) bacteria, and that same methane becomes food for methanotrophic (methane-consuming) bacteria that convert $CH_4$ to $CO_2$. Some of these thrive with oxygen, others without, and these bacteria also show considerable ingenuity when it comes to different types of metabolism. There are even methane consumers that convert methane to formaldehyde when they have access to oxygen.

Methane was first a gas for a few devoted scientists before it became relevant as a central component in natural gas. Methane has now become a hot topic in terms of the greenhouse effect and climate change. In the current debate, the question of climate change largely revolves around $CO_2$, with good reason. Yet when C bonds with 4H to form $CH_4$, an alliance is formed that is significantly more potent as a greenhouse gas. This has to do with bonds. In short, the single bonds in $CH_4$ help to capture more of the earth's infrared heat radiation than the double bonds in $CO_2$.[51] Heat absorption occurs due to molecular vibrations, and single bonds vibrate more than do stable double bonds. Methane is, therefore, feared because it can harbour some truly unpleasant surprises for the earth's climate, and because it is a more formidable greenhouse gas than $CO_2$. It is the combination of unpredictability and potency that transforms $CH_4$ into what is often known as a climatic 'wild card'.

Yet how much more potent is $CH_4$? Potency can be measured in terms of both intensity and durability, and methane is extremely intense but less durable than $CO_2$, which is a 'stayer' in the atmosphere. That makes the comparison more difficult. The issue also becomes muddled by the fact that some comparisons are made based on molecules, others on weight. On a molecular basis, $CH_4$ is forty times more potent as a greenhouse gas than $CO_2$. On a weight basis, however, the equation shifts by a factor of 2.75 in methane's favour, since methane's atomic weight is only 16 ($12 + 4 \times 1$), while $CO_2$ weighs in at 44 ($12 + 16 \times 2$). If we calculate by weight, methane proves to be 120 times more effective than $CO_2$. However, there is also the matter of durability. Many climate models, as well as the Intergovernmental Panel on Climate Change (IPCC), calculate the mean difference in warming potential over a hundred years as being that 1 kg of $CH_4$ is twenty to thirty times more potent than 1 kg of $CO_2$, because $CH_4$ breaks down more quickly in the atmosphere. This is a bit confusing, perhaps, but the point is that even if the contribution from $CO_2$ to the current warming trend is roughly double that of $CH_4$, we must consider ourselves lucky when it comes to the microbial methane consumers in ecosystems. Without these, the emission figures from the world's waters and wetlands would look very different and, climatically speaking, we would probably already be in the sauna.

In early winter, when the ice is clear, not mixed with snow and not too thick, it is easy to see the gas bubbles trapped beneath it. A couple of axe blows and waving a match over the hole yields a clear flame, and here we see the reason hydrocarbons are so popular. Methane, as we have seen, is a little brother among hydrocarbons, with only one meagre C ringed by 4H. Whereas $CO_2$ is the *result* of combustion, $CH_4$ and its closest relatives are a *source* of fire and energy. The next in the row of hydrocarbons is ethane ($C_2H_6$), then comes propane ($C_3H_6$), butane ($C_4H_{10}$), and so on, up to the 10 carbon atom called decane. (Decane, which has been shortened in modern English to dean, is also the term for the head of the university's faculty, a word based upon the number 10, 'decanus' – leader of ten soldiers.)

Methane is also a little brother when it comes to molecular size, but it is more energy-rich than its big brothers and is the main component in natural gas, with ethane figuring second, and other hydrocarbons like propane and butane comprising very modest portions. Methane is also known as firedamp, and in this context $CH_4$ also has a bad reputation, since high concentrations of $CH_4$ together with $O_2$ make for a sinister and explosive combination with the lives of thousands of miners on its conscience. Since $CH_4$ is odourless, in the absence of reliable measuring instruments it was impossible to have any warning of danger before the explosion, unless one had, for example, a canary in a cage. We have mentioned the canary before in the context of seabirds that act as whistleblowers when it comes to the poor state of the oceans' health. The real canaries, those that warned miners of dangerous conditions, lived in the mines and literally tumbled from their perches when concentrations of $CH_4$ (or the toxic, but odourless, gas carbon monoxide) became too high. At that point, the choice was either to vacate or establish better ventilation. Since that time, the coalminer's canary has come to represent an early warning sign, and the steadily increasing breadth of literature concerning climate effects on ecosystems is full of examples of such symbolic canaries among the world's flora and fauna.

The gas bubbling up from inland lakes and swamps is also known as swamp gas. It is mainly comprised of methane, but is also interspersed with other hydrocarbons and $CO_2$. It was also from the wetlands surrounding Italy's beautiful Lake Maggiore that Alessandro Volta, the man who also invented the battery and after whom the unit of electric potential is named, isolated and described methane as a gas in 1778. His inspiration came from the, if possible, even more productive and versatile Benjamin Franklin, who earlier had written an article about 'combustible air'.

Methane, it turned out, was mainly the product of a variety of bacteria's conversion of organic material in the absence of oxygen, but there are other sources of methane that do not involve life, such as serpentinization, where water, $CO_2$ and the mineral olivine produce $CH_4$ in geologically active regions.

## THE BENEFITS OF METHANE

It is, of course, carbon's role in the broad cycle, and therefore in the climate, that is this story's subject and centre of gravity, but $CH_4$ also demonstrates carbon's versatility. $CH_4$ first and foremost plays a key role as an energy supplier: 15 per cent of the world's energy supplies, and more than 22 per cent of its electricity, comes from natural gas. There is nothing so good that it is not bad for something: this also leads to significant $CO_2$ emissions. On the other hand, the emissions per energy unit produced are significantly lower for gas than for oil, not to mention coal.

Methane can also be synthetically produced in a number of ways, for example, by heating the right mixture of carbon and hydrogen or through the chemical reactions produced by an aluminium carbide and water ($Al_4C_3 + 12H_2O \rightarrow 4Al(OH)_3 + 3CH_4$) or an aluminium carbide and hydrochloric acid ($Al_4C_3 + 12HCl \rightarrow 4AlCl_3 + 3CH_4$). More important here is perhaps that methane forms the raw material for countless chemical products, many of them with a bad reputation: methanol ($CH_3OH$), formaldehyde ($CH_2O$), chloroform ($CH_3Cl$) and carbon tetrachloride ($CCl_4$), as well as some freons or CFCs. Yet methane is also a raw material for hydrogen, one of the potential 'pure' energy sources, for ammonia (as fertilizer), and for methanol, which itself is the basis for an endless array of petrochemical products.

Whereas ethane and propane are, respectively, $C_2H_6$ and $C_3H_8$, with their Cs connected by single bonds, ethene ($C_2H_4$) and propene ($C_3H_6$) have not just replaced an 'a' with an 'e' in their names, but are also equipped with double bonds, something that, in principle, paves the way for other types of products. Ethene and propene form the basis for the plastic materials polyethylene, polypropylene and PVC. Methanol (from methane) can be chemically converted, via a catalyst, to ethene and propene, and methane therefore potentially becomes the raw material for many of the same things we get from oil. Methane can even be made into diesel.

Yet what is most interesting, at least when viewed biologically, is the path from methane to food. Methane originates largely from the conversion of organic materials, but can this process be reversed? Can methane become protein without taking a detour through $CO_2$ and photosynthesis? One route is when methane-consuming bacteria are themselves consumed by single-celled protists or small zooplankton, thereby entering into the biological cycle. Still, there are greater visions for methane. In the 1990s the Norwegian companies Statoil and Hafslund founded the company Norferm to produce methane-based protein on a large scale. Their market was fish and animal feed, and in the quickly growing aquaculture industry it had become clear that importing fish protein from Chile in order to produce salmon protein in Norway could perhaps be sustained economically but not ecologically. The world's first bioprotein factory was opened in autumn 1998 with a yearly production capacity of 10,000 tons of methane-based protein.[52] Since proteins are amino acid polymers, they need an ammonium source. This is where methane comes in.

The factory's central unit of production was the methane-consumer bacteria *Methylococcus capsulatus*, which was fed a diet of methane gas, oxygen, ammonia and nutritive salts in large reactors at a temperature of 45°C. The bacteria then did what bacteria do best, multiplied, creating bacteria protein. At this temperature, and with good growing conditions, it does not take more than two to three hours for the bacteria to double, and double again, and double again. Anyone who has heard the legend of the chessboard inventor who, as a reward, demanded one grain of rice for the first square, two for the second, and so on, double for all sixty-four squares, has understood the principle behind bacteria's logarithmic growth. The 64th square requires $2^{64}$ rice grains, more rice than is found in the world.

The bacteria did not take it that far, but nonetheless the factory's production capacity was formidable. The end product contained 70 per cent protein by dry weight, an excellent nutrition source for fish and animals, and also undoubtedly for people. Norferm itself was not a commercial success, but others have taken the idea further, and I believe it is only a question of time (and demand) before methane becomes an important protein source – maybe not with *Methylococcus* as the production unit, but rather one of the many synthetic, specially designed bacteria pieced together from DNA molecules in the lab.

## ARCHAEA

It is nonetheless methane's role in the greater cycle that bears closer scrutiny here, and methane also underscores the fact that life also affects climate. Almost all methane produced is based on carbon that was once bound as $CO_2$ by photosynthesis. Most of photosynthesis's by-products can in some form or other be converted to methane, given the right conditions. Foremost among these is the absence of oxygen, and it is here that archaea enter the picture. Archaea ('ancient') bacteria are the earth's true aborigines, having dominated the globe for a couple of billion years before the oxygen build-up.[53] Many types of ancient bacteria are still with us today, many of them still dauntlessly producing methane in oxygen-free systems, from swamps and the muddy bottoms of lakes to the stomachs of termites and ruminants.

Archaea bacteria are bacteria in the sense that they lack a nucleus, which is the technical definition of a bacteria or prokaryote organism, but from an evolutionary perspective they are as distant from bacteria as we are from mushrooms. When it comes to the classification of living organisms, they are placed in a separate kingdom, the two other kingdoms being bacteria and eukaryotes (those whose cells contain nucleuses). As such, we are talking about a group that represents a separate growth from the evolutionary tree's root. It is, therefore, perfectly acceptable to call them simply archaea and drop the 'bacteria' label completely.

Archaea is a highly variable group with a downright masochistic fondness for extreme environments, whether it be scalding hot springs, acid baths, sulphur pools (often in combination), or the bizarre ecosystems you find in some deep sea areas where gases stream from the earth's depths through hydrothermal vents shaped like chimneys. Here the main type of hydrothermal vents are called 'black smokers' with temperatures around 350°c and black smoke composed of metal sulphides, as well as 'white smokers' with a somewhat lower temperature and white smoke composed of barium, calcium and silicon. These bizarre chimneys are home to their own communities of archaea, which form the basis for a particular kind of ecosystem with a distinctive wildlife. Archaea not only tolerate being roasted at temperatures in excess of 100°c, but have been found in tremendous health several kilometres down in Greenlandic ice, and they can be brought back to life after thousands of years encapsulated in Siberia's frozen tundra.

With good reason, archaea are often called extremophiles. One of the reasons for their extreme preferences is quite simply that these are the niches that remained after more modern organisms, above all oxygen-tolerant organisms, took over, but such systems are reminiscent of the conditions that existed almost 4 billion years ago, when our young planet was smoking, lifeless, oxygen-free and utterly uninviting for life as we know it. The fact that archaea seem to be relegated to the planet's environmental extremes, however, does not imply that their significance is likewise marginal. On the contrary, they are among the most numerous groups in the ocean, also appearing in the upper water layers (many tolerate oxygen quite well), and they are central players in the great biogeochemical cycles, such as the nitrogen cycle and the carbon cycle, especially when it comes to the quantity of $CH_4$ that alters the greenhouse gas window in the atmosphere.

Wetlands, including rice fields and inland lakes, are the dominant global methane source, but the world's ruminants, as well as its landfills, also contribute significantly. Indeed, even termites make a contribution to the atmosphere's methane content. It is, therefore, debatable whether we are dealing with natural or anthropogenic (human-induced) $CH_4$ emissions. Perhaps the correct answer is that we have literally gassed up a long line of natural emissions. Of course, it should be acknowledged that we in fact also contribute directly to methane emissions. Among the numerous microflora found in 'the human ecosystem', an assorted collection of methanogenic archaea inhabit the complex of the intestinal system. It is always amusing to remember that there are ten times more bacterial cells present in and on a person than our own cells. Of course, bacterial cells and archaea cells are so tiny that they are in the minority by weight compared to our own body cells, but together they still make up a few kilograms. The methanogenes contribute so much to gas production that it is inadvisable to experiment with a lit match back there – as some have done with unfortunate results.

The more than fifty known varieties of methanogenic archaea operate on a wide register when it comes to methane production, but all carry out their processes in the absence of oxygen. Some of them, the hydrogenotrophic archaea, actually use $CO_2$ to create methane ($CO_2 + 4H_2 \rightarrow CH_4 + 2H_2O$). It is this process that, given low temperature and low pressure, can form the frozen blocks of methane and water (ice), known as methane hydrates or clathrates, that keep isolated climate researchers awake at night – for reasons to which we will soon return. Within wetlands, guts and termites, it is usually acetate ($CH_3COO$-) that forms the basis for methane production. Acetate can originate from various sources, for example, from termites, tree cellulose or from foliage that breaks down into glucose, which is then converted into acetate. $CO_2$, furthermore, is also created in this process. The amount of $CO_2$ emitted by termites is not trivial, standing at an estimated 700 million tons per year, but this remains only a minuscule fraction of the global $CO_2$ output, so it is only on the methane market that termites make a difference. It is interesting to note that $CO_2$ reduction dominates on the ocean floor, whereas acetate reduction dominates in freshwater areas and wetlands.

The total annual methane emissions from 'natural' sources is calculated at 270 million tons or 0.27 gigatons of $CH_4$.[54] If the term 'natural' appears in quotation marks, it is only because it is debatable just how natural these emissions really are. Rice fields can hardly be termed natural, and we have contributed to a significant methane increase by flooding large swathes of land for use, for example, as power station reservoirs. Of these 270 million tons of methane, wetlands are responsible for 225, followed by termites with 20, whereas the ocean and gas hydrates are responsible for 15 and 10 million tons of $CH_4$ respectively.

It is reasonable to ask how it is possible to calculate overall methane emission from the unknown billions of termites. There are some 2,000 species of termites with highly variable methane emissions, and some species do not produce methane at all. Or how do we calculate emissions from the world's enormous and highly variable swampland areas, from countless lakes that come in all shapes and sizes – or from the world's oceans? Obviously, it cannot be done. What we can do is measure emissions from a select number of termite mounds and then multiply this figure with an estimate regarding the world's termite colonies. There is no need to point out that this approach is a blend of 'estimate' and 'guestimate', but at least it does provide a magnitude.

Even in my small forest lake it quickly became apparent that some of the funnels on the bottom had captured a lot of gas and others hardly any. The gas sometimes appears as large, individual bubbles, sometimes as tiny ones, and it is not evenly distributed across the lake bottom. So what about the ocean? In the large picture, the ocean plays a small role because conditions for methanogenic archaea are unfavorable. Conversely, the ocean could become a formidable methane source if it should for some reason start emitting its methane hydrates. Over time sensors and analysers for $CH_4$ have also been developed, giving some sense of the ocean's methane production, and most importantly these will warn if the ocean's methane hydrates begin to stir, even if it is difficult to imagine what in the world we could do about it.

It is easier to obtain realistic figures for anthropogenic methane emissions. The total is more than from natural sources, 330 million tons annually, with the primary and almost equal contributors being emissions from the energy sector (for example, from coal, oil and gas extraction) and emissions from the world's ruminant livestock. As we shall see, there are several reasons why beef leaves a high climatic footprint. Landfill, waste treatment and the burning of biomass are also contributors.

It is actually possible to distinguish between different sources of methane in the atmosphere because each source has a different isotope signature. Methane produced in ecosystems by archaea exhibit a different ratio between the isotopes 13C and 12C than methane leaking from gas pockets in the earth or on the seabed. Whereas 14C is a radioactive isotope that loses its spark after some tens of thousands of years, 13C is a stable (non-radioactive) isotope that has no similar expiration date.

The balance sheet for methane, however, also has a debit side. Here it is not photosynthesis, weathering and precipitation that come to our aid, but instead atmospheric breakdown that disposes of the estimated annual 550 million tons of $CH_4$. Some is also absorbed by the soil, where methane-consuming bacteria convert it to $CO_2$ (as also happened in my lake). We also contribute to a certain extent to reducing $CH_4$ emissions by trenching swamps and draining wetlands. As an initiative deliberately taken on behalf of the climate, however, this is a poor strategy since any gains are offset and then some by the increase in $CO_2$ emissions as the organic carbon stored there for millennia oxidizes. Aside from that, swamps and wetlands represent valuable biotopes, both in terms of their particular ecosystems and in terms of the role they play as flood control. Swamps and wetlands function like sponges that soak up water when there is an excess and release it in manageable quantities when conditions are dry.

Methane's total balance sheet, in terms of magnitude, looks something like this: 270 million tons (natural emission) + 330 million tons (human-induced emission) – 580 million tons (breakdown in the atmosphere and soil) = 20 million tons. For a corporate balance sheet, there would be little here to celebrate. Given the current situation, the optimal balance sheet for methane is not actually a balance, but rather 'red numbers' (actually, the blue numbers are the red ones in this balance sheet). Nonetheless, 97 per cent of annual methane emissions are absorbed, and that is good news. Even though the actual emission and absorption figures are uncertain, we have an idea regarding the net balance thanks to constant measurements taken by measuring stations the world over. These are largely the same measuring stations that keep tabs on $CO_2$ and the picture for both is the same: the $CO_2$ and $CH_4$ curves both point upward.[55]

As we now know, $CH_4$ has a significantly shorter atmospheric lifespan than $CO_2$. An average $CH_4$ molecule cannot count on lasting much longer than seven years, mainly because it is a favourite prey of hydroxyl radicals, OH. Because they are notoriously unstable, radicals are an intense substance in many chemical reactions. When high energy radiation, such as ultraviolet light, strikes organic material, chemical bonds can change, resulting in compounds with unpaired electrons, free radicals that can go berserk trying to correct their charge balance. That is one of the reasons we undergo gene and cellular damage from UV radiation (most of which, fortunately, is repaired by our cells' efficient machinery), but free radicals are also formed in other ways and are one of the reasons we age and wrinkle.

When, for example, water vapour in the atmosphere loses an H atom, the resulting $OH^{\cdot}$ radical (the dot after the OH symbolizes an unpaired electron) hunts feverishly for a new one, for example from $CH_4$. After meeting $CH_4$, the $OH^{\cdot}$ gets its wish and the result is $CH_4 + OH^{\cdot} \rightarrow CH_3^{\cdot} + H_2O$. Now it is methane that radicalizes, becoming an unstable methyl radical forced to hunt for another electron to fill its empty slot. After a number of intermediate reactions, the ultimate equation is $CH_4 + 2O_2 \rightarrow CO_2 + 2H_2O$. The balance has been reinstated, but meanwhile a methane molecule has been used up.

So where does the radical come from? Primarily from different chemical reactions in which, as mentioned above, ultraviolet light (UV) plays an important role. Hydrocarbons and oxidized nitrogen (usually termed $NO_X$) influence the formation of radicals, and both are components we supply to the atmosphere. We have already discussed hydrocarbons, but $NO_X$ is also another by-product of a society that exists by burning old hydrocarbons. Every single combustion process, whether occurring in a car motor, a coal-fired factory or an oil fire, transforms some of the atmosphere's free and non-reactive $N_2$ to $NO_X$. Much can be said on this subject, and we will certainly return to it, but the point here is that the significant quantities of $NO_X$ that we add to the atmosphere include the radicals $NO^{\cdot}$ and $NO_2^{\cdot}$.

Radical formations are rather like a game of musical chairs where the point is to snatch electrons from others, something that produces new molecules, which in turn are left with unpaired electrons (thereby forming radicals). How all this will ultimately affect $CH_4$ remains undetermined, but there are many indications that increased radical contribution from $NO_X$ can help reduce methane's lifespan. In the same way that increased nitrogen emissions can fertilize forests and oceans, thereby helping bind more atmospheric nitrogen, $NO_X$ can help keep atmospheric methane in check. Here we can also establish that there is nothing so bad that it is not good for something.

Like so much else, the $CH_4$ concentration in the atmosphere is a mass balance ($CH_4$ in minus $CH_4$ out), and is primarily regulated by the factors that produce methane in wetlands (favourable conditions for the type of bacteria), the bacteria that 'consume' methane (other types of bacteria), and the amount of free radicals present in the atmosphere. In contrast to $CO_2$, reduced $CH_4$ emissions will within a short time span lead to atmospheric decreases. So why is there in fact more methane? Has something gassed up the archaea? Obviously, they now have more livestock and rice fields in which to frolic, but the changes are also a result of more or less involuntary emissions from the fossil fuel industry. It is no exaggeration to say that our future hangs on the carbon cycle and how we approach it, so the time has come to take a closer look at what we know.

# PART II

# THE C IN CYCLE

Perhaps the photosynthesis equation deserves to be termed the world's most important biochemical reaction. I would cast my vote for it, at any rate, even though there are many other strong candidates. Similarly, it seems natural to award DNA the status of life's most important molecule – at the very least, that helixed genetic material is the world's most famous and iconic scientific symbol, containing the innermost core of life and an individual's genetic fate. When it comes to the world's most important scientific figure, the competition gets even fiercer, but the way things stand today I would vote for the Keeling Curve, which shows $CO_2$ concentrations taken from the measuring station atop Hawaii's Mt Mauna Loa.[1]

## THE KEELING CURVE

Charles David Keeling (1928–2005), like many other great names in carbon's history, ended up in chemistry by chance. At the age of twenty Keeling began specializing in what was then a hot field within chemistry: polymer chemistry. Keeling finished his studies in 1953 and could easily have settled for a secure job within the blooming plastics industry. Keeling was an outdoorsman by

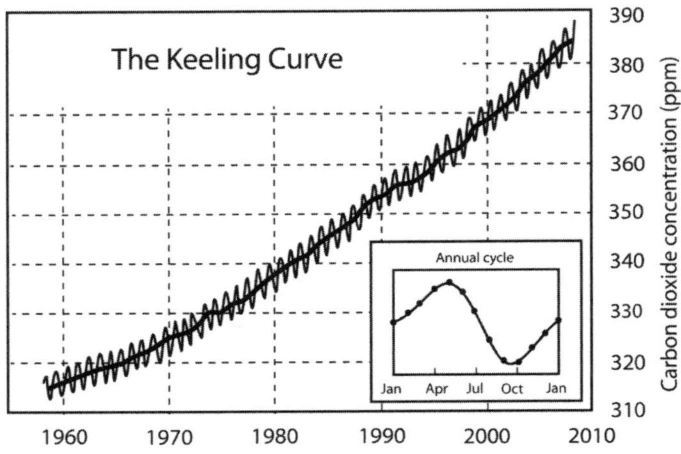


The Keeling Curve showing the gradual increase in atmospheric $CO_2$ at the monitoring station in Muana Loa, Hawaii.

nature, however, and did not feel called to a life within industry. Instead he embarked on a more uncertain career path by combining post-doctorate research in geochemistry at the California Institute of Technology (Caltech) with his mountain forays. It was his Caltech adviser who first raised the idea of exploring whether it was possible to calculate groundwater's carbonate concentration upon the assumption that water stood in equilibrium with $CaCO_3$ and atmospheric $CO_2$. Understanding carbon's pathways through air and water was also central to geochemistry back then, even though such questions were not motivated by potential climate changes.

Keeling accepted the challenge to test the idea, which required an instrument capable of taking precise $CO_2$ measurements. One problem was that in the 1950s no one knew exactly what the $CO_2$ concentration was and a sufficiently precise measuring instrument did not exist. Most measurements pinpointed the amount as somewhere around 300 ppm (parts per million; that is, 0.03 per cent), but estimates varied from 250 to 550 ppm. Keeling's initial tests showed that the result clearly depended upon human proximity: our exhalation was enough to significantly impact measured $CO_2$. Concentrations depended not only on *what* was taking measurements, but on *where* those measurements were taken. He also discovered that $CO_2$ concentrations were higher at night than during the day and realized that vegetation and other near-ground conditions notably influenced the results. Eventually the original question became less relevant, and Keeling instead became almost obsessed with collecting reliable data on $CO_2$.

Fortunately for Keeling, the International Geophysics Year of 1957–8 brought with it the momentous ambition to measure everything capable of being measured within the realm of geophysics. Healthy grants followed suit, particularly because in that same year the Soviet Union started the space race by launching Sputnik I. 'Sputnik shock' shook the USA, which quickly ramped up research grants, especially for anything to do with space and space technology. As a direct result, NASA was established in 1958 and the agency flourished on generous allocations, while some trickled down to other projects. Crediting Sputnik and the Soviets for the Keeling Curve would be an exaggeration, but their indirect contribution seems obvious.

Rumours of Keeling's thoroughness and analytical capacity reached Harry Wexler, Chief of the Scientific Services division of the U.S. Weather Bureau in Washington, DC. Wexler was weighing plans for several $CO_2$ measuring stations, one of them atop the 3,400-m-high Mt Mauna Loa in Hawaii, far removed from all human and near-ground influences. Wexler invited Keeling to discuss the possibilities and limitations of measuring $CO_2$, a topic few knew better than Keeling – at least, he was familiar with the limitations of the old method he himself had used, a quicksilver manometer.

The greenhouse effect had long been recognized, though not generally considered problematic in the 1950s, so it was well known that $CO_2$ absorbed long-wave infrared radiation. Keeling had the idea of using this principle to achieve exact measurements of $CO_2$ concentrations. The greenhouse effect used to measure its very cause! Wexler bought the idea immediately and the next day offered Keeling a job.

Keeling ordered two instruments built, each equipped with a sensor that measured infrared light reduction, itself proportional to the quantity of $CO_2$ present. One instrument was sent to Antarctica, the other to the top of Mauna Loa, where it would continue to work from 1958 until it was retired in 2006. The original plan was to take measurements for the Geophysical Year of 1958, with maybe another measurement taken twenty years later, simply to determine whether the concentration stayed constant. In May 1958 the maximum was measured at 315 ppm, though in late autumn the results sank to a minimum of 310 ppm. There were some operational problems, however, and it was not clear whether this 5 ppm variation was real or due to the instrument. As a result, it was necessary to extend the measuring period until 1959. That year everything worked perfectly and the result confirmed the 1958 measurements. The apex during spring was correct, right before the trees grew their leaves and fired up their $CO_2$-absorbing photosynthesis machinery, while the low point during the autumn was the result of the summertime carbon sequestration by vegetation in the sea and on land. Keeling himself described the fluctuations in the following terms: 'We were witnessing for the first time nature's withdrawing $CO_2$ from the air for plant growth during summer and returning it each succeeding winter.' If one desires a 'mother Earth' perspective, it is here that one could actually see the earth inhaling and exhaling, albeit not in synchronicity between the northern and southern hemispheres. Whereas these seasonal fluctuations would have once been written off as measuring noise, Keeling's efforts had resulted in a precision instrument that provided a glance into how the balance between photosynthesis and respiration actually produces measurable fluctuations in atmospheric $CO_2$.

In 1961, however, the powers that be felt these answers were enough. The Geophysical Year had come and gone, the astonishing grants were used up, and there was no obvious reason to spend more money on monitoring a trivial gas. It was obvious to Keeling, however, that the measurements had to continue, particularly because the instrument's very precision had alerted him to a peculiar increase in $CO_2$ concentrations that could not be written off as an instrumental defect. In 1960, of course, the increase was only a modest 0.7 ppm per year, whereas the average increase across the decade from 2004 to 2014 was 2.07 ppm per year. In answer to suggestions that he stop taking measurements, Keeling issued warnings of the possible climatic effects the increased $CO_2$ concentrations might herald. After some serious lobbying, he scraped together barely enough grant money to ensure the monitoring could continue, at least for a few more years.

In 1963 he met with a number of other experts to discuss the significance of the peculiar increase in $CO_2$. They concluded in their report that, over the course of the next century, a potential doubling of atmospheric $CO_2$ could cause average global temperatures to increase by 4°C. Roughly fifty years later, we cannot say that these warnings prompted any major social changes, but perhaps they allowed Keeling to carry on. In the following years $CO_2$ continued to increase, and the Keeling Curve has become the iconic symbol of Earth's – the patient's – rising fever. Keeling continued to publish his findings and warnings, which proved much of the basis for later climate research. He also realized that, given the continuous gas exchange between the atmosphere and the ocean, there would also be a steady increase in oceanic $CO_2$ concentration, resulting, for example, in acidification.

Ultimately Keeling became a kind of godfather for greenhouse gas research. He was the first to submit clear evidence that the carbon cycle was in distress. The Mauna Loa station was followed by measuring stations the world over, all of which showed the same upward progression. The first record that we had exceeded the symbolic limit of 400 ppm was actually made in 2013 at the Norwegian observatory atop Mt Zeppelin, which is close to Ny Ålesund in the Svalbard Islands.

MAUNA LOA is an old volcano, occasionally reminding us of this fact by expelling a little extra $CO_2$ from the depths. Keeling, of course, was conscious of this. As a rule, it was a simple matter to weed out these isolated peaks from the steadily rising curve, but it also illustrated a returning objection from those who claimed the $CO_2$ increase came from places other than coal factories, industry, aircraft and automobile traffic, and so on. Nature itself, volcanic gas leakage, for example, was the cause. This was a strange objection given that it was always a simple equation to show by magnitude the amount of $CO_2$ generated by the combined consumption of fossil fuels and cement production. It was obvious that steadily increasing usage of fossil fuels must create steadily increasing quantities of $CO_2$ and that all that extra $CO_2$ must end up in the atmosphere. What was not so simple was calculating the $CO_2$ contribution from deforestation, fires and landscape changes in general, but these factors all contributed substantially on top of coal, oil and gas.

In fact, what was astounding was not that $CO_2$ was increasing, but that it was not increasing *more*. Roughly half of what we emit accumulates in the atmosphere, so the question was not the amount of $CO_2$ nature contributed to the atmosphere, but rather how much of our own emissions was being controlled by processes within ecosystems. Keeling's Curve should have been substantially steeper. On a very long time horizon, on a 100,000-year timescale, $CO_2$ will be sequestered by mountain weathering. As such, there had to be other mechanisms, undoubtedly photosynthesis as a primary one. Did this sequestration, however, take place in the ocean or on land as well? Keeling realized that we needed accurate measurements of both $CO_2$ in water and $O_2$

in the atmosphere. As such, he started monitoring the ocean's $CO_2$ concentrations, which, sure enough, turned out to be rising, while the pH sank. The ocean was becoming more acidic.

As to changes in the atmosphere's $O_2$ content, that was a subject Keeling took up with his son, Ralph. Because $CO_2$ originates through combustion processes, there should be a reduction of $O_2$ in the atmosphere according to a stoichiometric relationship regarding the amount of produced $CO_2$ or combusted C (that is, two oxygen atoms disappear per $CO_2$ molecule produced). Ralph Keeling inherited both his father's interest and talent. Whereas his father developed a sensitive analytical instrument for $CO_2$, Ralph developed an even more sensitive instrument for measuring atmospheric $O_2$, with which he clearly demonstrated that the decrease in $O_2$ concentration corresponded to the $CO_2$ increase.

This does not imply that the atmosphere's oxygen content is in danger of being used up. While for every molecule in the atmosphere there are around 400 ppm of $CO_2$ molecules, there is a much richer supply of oxygen, nearly 210,000 ppm. Even if all known deposits of coal, oil and gas were combusted, it would only consume 1 per cent of the atmosphere's oxygen supply, so that is not where the problem lies. Ralph later took over 'the family business', assuming responsibility for the Mauna Loa observatory.

In 1988 constant measurements of methane were initiated on Mauna Loa and it quickly became clear that $CH_4$ levels had also crept up. Supplemented with ice core measurements, it was shown that $CH_4$ in the atmosphere had climbed from a relatively stable level of around 700 ppb in pre-industrial times to around 1,800 ppb today, that is, an increase of 250 per cent. Note here the use of *ppb*, parts per billion, to indicate quantity. And there is a further source of confusion. Even in serious articles and books, there is a tendency to make a mistake here, but methane is even rarer in the atmosphere than $CO_2$. Whereas there are 400 ppm of $CO_2$, there is only 1.8 ppm of $CH_4$. This 'only' is more than enough, however, since today methane contributes to around 25 per cent of the greenhouse effect.

Whereas the $CO_2$ increase has been steady, predictable and comes from some obvious sources, $CH_4$ is more unpredictable, probably because it has more sources. Methane increased quickly until 2000, levelling out until 2006, but is now on the rise again.

## A SWEDE BEFORE HIS TIME

It was Keeling who demonstrated that humanity had reached the point of impacting both weather and climate. By the time Keeling began his measurements, however, the idea that $CO_2$ could affect the climate had already been known for more than sixty years. We must turn again to Sweden, and more particularly to Svante Arrhenius, who won the 1903 Nobel Prize for Chemistry and served as the director of the Nobel Institute from 1905 until his death in 1927.[2] In 1896 Arrhenius estimated the consequences of a doubled $CO_2$ concentration, a scenario that, it should be noted, was then regarded as purely theoretical. His estimates were more a means to illustrate the importance of $CO_2$ for maintaining a liveable climate on earth, but his quick calculations for what a doubled $CO_2$ content might imply did not depart much from Keeling's conclusion sixty years later, or from the International Panel on Climate Change's (IPCC) 'business as usual' scenario a hundred years later.

In his 1906 book *Worlds in the Making*, which was published in numerous editions and was immensely popular among progressive circles in Sweden, even though it was by no means light reading, Arrhenius contextualized his calculation using 300 ppm $CO_2$ as a springboard: 'I have calculated that if the atmosphere were deprived of all its carbonic acid – of which it contains only 0.03 per cent by volume – the temperature of the earth's surface would fall by about 21°.' He further reasoned that this would reduce the earth's water vapour content, leading to additional cooling. Can $CO_2$ fluctuations thereby explain historical climate fluctuations and perhaps predict the future?

> If the quantity of carbonic acid in the air should sink to one-half its present percentage, the temperature would fall by about 4°; a diminution to one-quarter would reduce the temperature by 8°. On the other hand, any doubling of the percentage of carbon dioxide in the air would raise the temperature of the earth's surface by 4° . . .
>
> The question, however, is whether any such temperature fluctuations have really been observed on the surface of the earth. The geologists would answer: yes.

Arrhenius also takes up the subject of human emissions, not without a certain anxiety: 'The actual percentage of carbonic acid in the air is so insignificant that the annual combustion of coal, which has now (1904) risen to about 900 million tons and is rapidly increasing, carries about one-seven-hundredth part of its percentage of carbon dioxide to the atmosphere.'

Arrhenius indicated that the growth of coal consumption had been formidable, more than doubling from 1890 to 1907, and even if the ocean eventually absorbed much of the resulting $CO_2$, the increased consumption would still result in an increase of $CO_2$ in the atmosphere. He also explained that the other slow processes that counteract $CO_2$ accumulation in the atmosphere are weathering processes that bind $CO_2$ and, referring to 'the famous chemist Liebig', he discusses the link between the slow geochemical cycle, which is governed by weathering and chemical precipitation, with the biological cycle. Increased $CO_2$ will also stimulate plant growth, and more plants (and plant roots) will increase weathering, which in turn will liberate important minerals for plant growth, especially phosphorus. This process will itself further stimulate plant growth, again contributing to an increased absorption of $CO_2$. Since a large amount of plant material is cellulose, which contains 40 per cent carbon, Arrhenius calculated that 'the actual annual carbon production by plants would amount to 13,000 million tons – i.e., not quite fifteen times more than the consumption of coal.'

In short, Arrhenius, who was an impressive visionary, managed to outline in a few pages the central insights into the carbon cycle, while also citing emissions from fossil energy sources as having a potential impact on the climate. His remarks also anticipated many of the mechanisms for global thermostat regulation that are central to James Lovelock's Gaia theory, which we will discuss later. Arrhenius is not entirely accurate when it comes to estimates surrounding vegetation's annual carbon sequestration, which is only to be expected, but his suggestion regarding the temperature effects of a doubled $CO_2$ quantity is astonishingly close to the IPCC's predictions. It is also worth remarking that Arrhenius's estimate was the same as Keeling's baseline sixty years later: 300 ppm $CO_2$ in the atmosphere. This was no mean estimate, but it was also not enough to say what was actually happening.

While mentioning early Scandinavian researchers, we should again acknowledge Sweden's Carl Linnaeus who, though he lacked Arrhenius's analytical abilities, made the most of his intuition. Linnaeus was perhaps the first person to describe life cycles and food chains, which he did with impressive insight. He also illustrated the short carbon cycle in a precise, if rather morbid, way typical of the mid-eighteenth century, by explaining how human heads currently decomposing in the cemetery would become the soil in which cabbage heads might be grown and then eaten by humans.

Linnaeus was not exactly quick to credit colleagues, but Arrhenius never hid the fact that he was standing on others' shoulders, such as those of James Tyndall.[3] In 1859, just as Darwin was creating a furore with *On the Origin of Species* – and

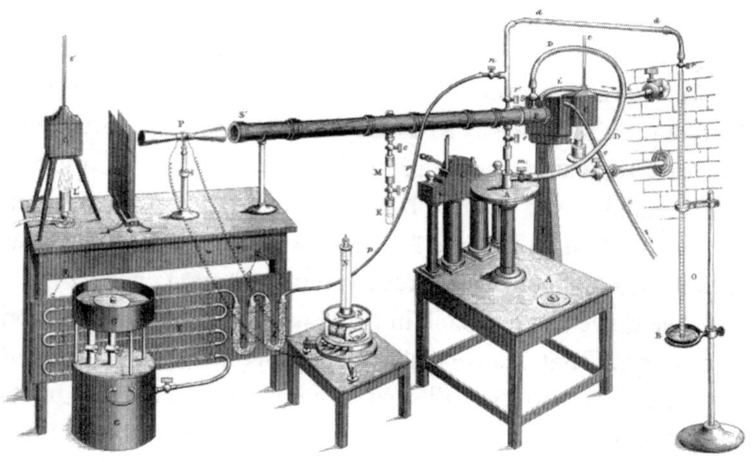

James Tyndall's home-made apparatus, which led to the basic discovery of the ability of certain gases to absorb heat radiation, and thus cause the greenhouse effect.

a century before Keeling enjoyed his first, uninterrupted year of observation on Mauna Loa – the Irish physicist James Tyndall conducted a series of impressive experiments that formed the basis for Keeling's precision instrument for measuring $CO_2$. Tyndall's home-made but elegant layout for measuring how different wavelengths of radiation behave in the air, prompted him to conclude that warmth in the earth's atmosphere could be explained by the ability certain gases had to absorb infrared radiation or thermal radiation, that is to say, the long-wave and invisible portions of the light spectrum.

Tyndall was able to measure the infrared light absorption ability of gases like $CO_2$, $CH_4$ (methane), $N_2$ (nitrogen gas), $O_2$ (oxygen), $O_3$ (ozone) and water vapour. He noted that while $O_2$ and $N_2$ do not, practically speaking, absorb long-wave radiation, other gases, including water vapour, had many areas of absorption. That meant that energy or warmth in areas of long-wave back radiation from the earth's surface were 'trapped' by these gases. The quantity of these gases, therefore, would logically affect earth's temperature. The fact that methane also played a central role in warming meant that C had stock in two major greenhouse gas molecules, $CO_2$ and $CH_4$. With these insights, Tyndall provided the key to an old question: how are earth temperatures regulated? It was a significant enough contribution to land Tyndall in the science Hall of Fame. In 2000 the Tyndall Centre for Climate Change Research was created, one of Great Britain's leading centres for research into climate change.

If we delve further back in time, we find that this question is something that Jean-Baptiste Joseph Fourier asked himself forty years prior to Tyndall's experiments. When the sun heats the earth, why does the earth not continue to absorb heat until it finally becomes as hot as the sun itself, or alternatively lead to a kind of equalization between a cold and a hot object as toward an equilibrium temperature for both? Fourier made the entirely correct assumption that the earth must emit long-wave radiation invisible to the human eye. Parenthetically, we should remark that Fourier, like so many early masters in the natural sciences, was an all-rounder. He led a rebellious youth and was arrested for revolutionary activities. He also served Napoleon as scientific adviser during the Egyptian campaign, but is actually most famous as a mathematician for his analysis of periodic functions, known as the Fourier analysis, not to mention his contributions to various mathematical and probability theories.

Fourier's springboard became his attempt to calculate the heat effect of the earth's radiation balance, and in 1820, following painstaking calculations, he reached the somewhat surprising conclusion that the earth should have been significantly colder than it was, far below freezing in fact. He launched several theories to explain this paradox, one of them being that the atmosphere must have a form of isolating effect that keeps temperatures on earth liveable. At this point, we will not go further back in history, but can perhaps call Fourier the spiritual source of greenhouse gas theory.

Fourier, Tyndall and Arrhenius built step-by-step the theoretical staircases leading to Keeling, who in turn supplied the theory with actual measurements in the form of the Keeling Curve's increasingly steeper slope. These four were not alone and others also helped to link gases, particularly $CO_2$, to a potential warming effect. Some regarded the idea as indisputably beneficial, something that would secure increased crop production and protect us from approaching ice ages (an argument we can still occasionally find today). Others believed that, while warming potential was real enough, the ocean also contained many times as much $CO_2$ as the atmosphere, so any excess $CO_2$ would be absorbed there, meaning the atmosphere would remain stable. Up until 1960 one could certainly hold that belief, but from then on it was clear something was fermenting, something whose cause lay in the fact that a rapidly growing population with rapidly growing energy consumption nursed that consumption by conveying carbon, which had taken millions of years to sequester, and which had also been out of circulation for millions of years, back into the atmosphere and biosphere at record speed. The key to further insights here was found in the carbon cycle.

## ONE OR MANY CYCLES?

It is, of course, a gross oversimplification to talk about *the carbon cycle* as if it were a single cycle. There are countless cycles taking place on different scales in time and space. Some happen within cells, such as carbon fixation through the photosynthesis cycle, while others come about through cellular respiration in the Krebs cycle, which gives us energy by combusting carbohydrates, fats and proteins into $CO_2$. Some cycles involve interactions among several different species and might take place on a timescale of hours or days. Some cycles happen on a yearly basis; some take a century or are spread over a timespan of several hundred thousand years. All these cycles are linked, though, a fact that does nothing to simplify the matter. If someone claimed to thoroughly understand the carbon cycle, you could allow yourself an ironic smile. It is notoriously difficult to pinpoint what is driving what within the carbon cycle or cycles, as well as how this turns around and affects the climate. Yet, inscrutable or not, carbon's many pathways will decide our collective fate for the foreseeable future. With these words, I am discounting good-sized meteorites, gamma ray bursts, a blowout in the pressure chamber beneath Yellowstone, and other things of that nature, given that it makes the most sense to concentrate on what you can actually do something about.

When it comes to the carbon cycle, one might think that the narrower the focus, the more one would understand. Yet as the endless dietary debates illustrate, even when it comes to our own carbon cycle there are plenty of viewpoints and unknowns in the equation. Each one of us leaves a footprint on the global carbon cycle that is significantly greater than we are individually capable of combusting throughout our life with our own respiration, but for now we will let that lie and turn our focus inward.

No matter what it is, everything we eat contains carbon, which is typically divided into three main types: carbohydrates, fats and proteins. As cellular fuel these three groups have different roles to play. Carbohydrates (which, incidentally, were once called 'coal hydrates') are sugars and fire cells most easily. They offer plenty of energy by weight unit, but little else. Whereas the simple sugars, the monosaccharides and disaccharides (what we usually call 'sugar'), are effective fuels, polysaccharides like starch and cellulose are difficult to burn. Fat is also a relatively high-octane fuel. However, fat's primary function is storing energy against future need (as well as insulation). And as the ongoing health debates perpetually remind us, fat comes in many varieties. Saturated, unsaturated and polyunsaturated fats all have different traits and roles, but we should not get distracted. Proteins also store energy but occupy a completely different and key function as building blocks in muscle tissue, neurotransmitters and other things.

All food holds energy in the form of organic carbon, which is actually solar energy plants have generously converted into something from which we animals can benefit. The first step is for this carbon, packaged in its different ways, to enter the body: it will be eaten, chewed and swallowed. At that point the digestive system will decide whether the food is something that can and should be incorporated into the body. Typically, an assorted selection of digestive enzymes will strive for the maximum carbon absorption possible, but some carbon is beyond reach. Cellulose is the main component of plant cellular walls and makes up a large portion of trees in general, but it is also the main component in cotton, linen and plant fibres. The polymers in cellulose, which can be thousands of repetitions of $C_6H_{10}O_5$, are so tough for the enzyme apparatus in the intestines to tackle that the only things capable of performing this feat are bacteria and some single-celled organisms. Cellulose has many different uses, but as food it is seen as slim fare. Those wishing to digest cellulose, whether it be termites or other creatures, must ally themselves with intestinal symbionts that are up to the task.

All the carbon assimilated through the intestinal walls will largely be broken down into simpler components that are redirected and re-emerge as our body. No small percentage goes to combustion within mitochondria, whose task it is to keep our temperature up, as well as to supply energy for the countless things that require some form of energy or other. The final product of this combustion is the $CO_2$ we exhale. If there is any carbon left after all the necessities have been covered, it is deposited in the bank as fat. This has been a smart and mandatory security measure for most of our history, since as a rule we have suffered from a chronic energy shortage broken by ecstatic, gluttonous moments after the buffalo was felled. It was a wise move back then to set aside some fat on the thighs for a 'rainy day'. Some bodies are quite good at this, others not so much. Although today storing fat does not seem so wise, lean times can always come again, which will return the strategy to its dignity and honour.

In principle, the calculation is a simple one involving pluses and minuses, most easily set up using carbon as a common unit, since carbon enters into all the equation's components: carbon consumed minus carbon eliminated (not absorbed) minus carbon respirated ($CO_2$, that is) equals carbon accumulated (usually in muscles, but at times in Michelin rings). The balance can be changed by changing intake or expulsion, which is essentially what thousands of slimming tips will tell you. Ultimately the great cycles are a question of accounting and balance, but here the question becomes a little different. Here we do eat away some stores, but what we do *not* want to see happen is a situation where more is going out than in. That is the basis of carbon cycles, however small or large they may be. The principle seems simple, but the reality is complex.

A person's carbon budget does not fundamentally differ from that of other animals, and since the non-vegetarians among us typically consume something that has consumed something else, we form the last link in a series of such connected carbon budgets. We are at the top of the food chain, and if the food chain is long enough a substantial amount of carbon will have been deposited at the bottom to ensure enough is left over to keep a person going. It is now time to expand the carbon budget from cells and individuals to ecosystems.

## PYRAMIDS, CHAINS AND WEBS

Biology as a discipline was long devoted to describing species, preferably exotic ones from exotic places, in a naturalistic kind of stamp collecting that did not see much action between Aristotle and Carl Linnaeus. For Linnaeus, nature was a static system with species created all at once, although he did have a sense for the cycle whereby plants grew from the earth and were eaten by animals, who in their turn returned to the earth.

Roughly a century after Linnaeus' straightforward description of the simplest ecological carbon cycle imaginable, Charles Darwin tackled the subject in more detail. Evolution was not just about besting equals within a single species, but about all manner of adaptations within the ecosystem. Plants and animals participated in mutual adaptation, and Darwin's discussion of why clover crops in the English countryside vary is an elegant description of food chains. Crops were, of course, climate dependent, but the most important factor was pollinating insects, especially bumblebees. In the years bumblebees abounded, effective pollination took place, which meant good yields. In the years and regions where voles abounded, there were fewer bumblebees, since voles had the unfortunate habit of raiding bumblebee nests. More clover would grow right next to towns because there were cats to keep the vole population down. If we translate the actors in this drama to carbon-units, a portion of a cycle emerges that naturally can be made endlessly complex by drawing in all the actors, from bacteria to raptors and humans.

It was another British scientist, Charles Elton, who formalized the cycle of food chains or food webs.[4] In 1921 Elton, who was then a 21-year-old zoology student, was given the chance to participate in an expedition to Bjørnøya (Bear Island) and Svalbard with his supervisor Julian Huxley (the grandson of Darwin's standard-bearer and close ally Thomas Henry Huxley). Elton, together with the young botanist Victor Summerhayes, studied the simple ecosystem of Bjørnøya, focusing on the ecosystem's nitrogen cycle instead of the carbon cycle, but the principle is the same. Through detailed and elegant analysis, they demonstrated how all the actors in the ecosystem are intimately tied together. They began with free atmospheric nitrogen ($N_2$), which first was assimilated by bacteria before making its way through plants and animals into the soil and water, from the microscopic mites present in the sparse soil, and the even smaller rotifers and algae present in the water, via insects, an assorted variety of birds, drawing in seals, arctic foxes, polar bears and others. This made for an impressive and intricate network of interactions and organisms that are all connected through the nitrogen cycle. The same observation would also apply to the carbon cycle. It was one thing, however, to draw arrows between all the producers or consumers in the various hierarchies in the food web, but how could all this be quantified? Is it even possible to append numbers here, preferably in carbon units?

In 1942 the well-respected journal *Ecology* published an article entitled 'The Trophic-dynamic Aspect of Ecology.'[5] Unfortunately the author, Raymond Laurel Lindeman, did not live to see his work published, as he had died of a rare type of pneumonia earlier that year at the age of 27. The article was part of Lindeman's doctoral thesis and was based on his studies at Cedar Bog in Minnesota, a lake not unlike mine, and I can well imagine young Raymond sitting at the water's edge on a sun-filled summer's day, watching dragonflies hover over the reeds and wondering what was going on in the dark depths. And he found an answer. By studying producers and consumers, he constructed a complete energy budget inside a complicated food web. Lindeman measured energy transfer, and although our unit is carbon, the principle is the same. He calculated how effectively energy, or carbon, is transferred between *trophic levels*, trophic level 1 being plants, level 2 being plant-eaters, level 3 being whatever eats herbivores, and so on. Lindeman's rule of thumb became that 10 per cent of available energy from one level is transferred to the next. In other words, 100 kg of plant carbon will yield 10 kg of hare carbon, which yields 1 kg of fox carbon. The result is trophic pyramids, which narrow at the top because much of what is consumed is never utilized and much goes up in smoke, or $CO_2$.

As such, every trophic level through which carbon travels exacts a kind of carbon tax, meaning the ecosystem's carbon cycle is essentially as follows: plants bind carbon, and as the bound carbon travels through the food chain it becomes increasingly diluted. With every link, some vanishes into the large pot of dead organic carbon (what is not digested) and some into $CO_2$. Warm-blooded animals will pay a higher carbon tax than cold-blooded, because maintaining a stable temperature means heating costs taken in the form of organic carbon. Humans, moreover, expend energy not just to maintain 37°C, but to maintain our energy-craving brain, which demands no less than 20 per cent of our energy budget. This means there is ultimately nothing remaining for another trophic level. Elton and Lindeman conducted their studies at small lakes, which, owing to their isolation, are well suited to illuminate the cycle. That brings us back from Elton and Lindeman's lake to my own forest tarn.

After painstaking measurements and calculations, it is now time to do the accounting. What do we have in the bank? Owing to the variety of securities, currencies and denominations we are dealing with, we will assign carbon as the common currency here and micrograms (μg) per litre as the common measuring unit. The textbook example of a lake food chain has a platform of planktonic algae that forms the food chain's basis. As we move up the pyramid via algae-grazing zooplankton to plankton-eating fish, with perhaps predatory fish at the top, the pyramid narrows. There is perhaps a certain justice to the fact that those enthroned at the top, those that live off others, are quantitatively inferior in their own right.

My lake did not entirely follow the textbook example, as is usually the case with complex reality. The total main deposit in the water's carbon bank is loose, organic carbon: brownish, humic compounds that are the residual products of the forest, swamp and soil around the lake. As it turns out, there is more than ten times of this dissolved, organic carbon than the second largest deposit, inorganic carbon, which is mostly $CO_2$. In third place we find dead particles of various kinds. Then comes zooplankton, bacteria and phytoplankton, which make up the living components. A reliable estimate for fish and bottom-dwelling animals was quite simply impossible to give, but even a generous estimate for these groups contributes only modestly to the quantity of living carbon present. As such, the relationship of dead carbon to living is almost 40:1. Therefore, the main basis for the food chain in this lake is not plants (algae), which amount to only a modest 1 per cent of all the carbon in the lake's water mass.

In short, our studies yielded two central insights. There is an enormous carbon pump where carbon is taken up by land vegetation during photosynthesis. Most of the carbon proceeds resulting from photosynthetic labour is stored for shorter or longer terms in trees or soil, but some runs into water. Here this organic carbon will undergo photosynthesis's inverse

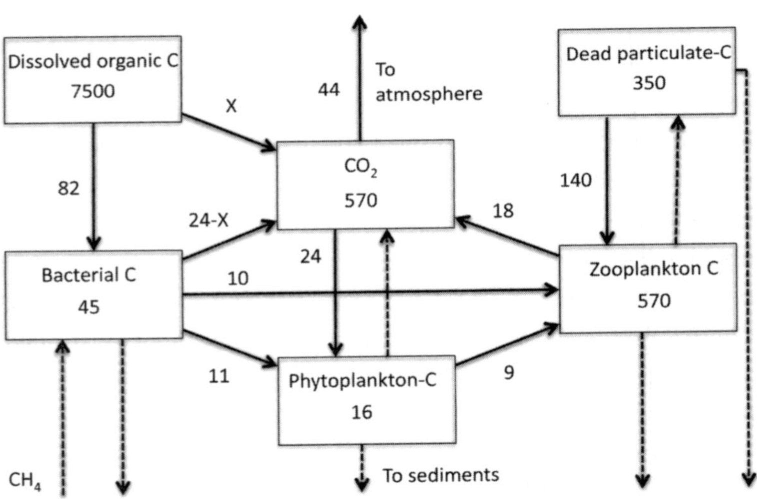


The carbon cycle – much simplified here – on a summer's day in a forest lake. The boxes display the carbon pools (microgram of carbon per litre), while arrows represent fluxes (microgram of carbon per litre per day). This is essentially the basic findings from our experiment, and it was obtained by use of the isotope 14C.

reaction, respiration. The organic carbon produced by land plants becomes an energy source ('food') for bacteria that in turn enter a food chain. Many things out there consume bacteria, some of which belong to a combined class that actually shifts between conducting photosynthesis like a proper plant and then devouring bacteria like an animal. Indeed, broadly speaking, the whole lake behaves like an animal, 'consuming' plant carbon and exhaling $CO_2$. One side effect of this strange cycle, which we believe applies to most lakes in the boreal coniferous forest belt, is oxygen-poor deep water. As we have seen, this fact, combined with a hefty portion of organic carbon, provides the foundation for quite substantial methane production. Bacteria transforms some of this methane into $CO_2$, but much finds its way into the atmosphere, thereby making its minuscule contribution to the greenhouse effect. In many respects, this kind of forest lake acts as a valve for returning $CO_2$ that is bound in forests and swamps, and represents an extreme variant of water's carbon cycle.[6]

I often jog along a river with beautiful waterfalls close to where I live. Periods of heavy rain turn the waterfalls brown with humus, the product of the forest's photosynthesis: the dense, carbon-rich and extremely complex residues of lignin and cellulose. They capture light and advertise their presence by their unmistakable brown colour (like when you dip a teabag into hot water). A tiny portion of river water originates from the small stream flowing from the forest lake. As it travels to the fjord, the river will discard as much as half of this loose, brownish carbon. This will return to the atmosphere, ready for a new turnaround, perhaps blowing northward, perhaps entering the stomata of a forest fir, where in a few years it will once again land in the river. The river water also carries along weathering products from soil and mountains, calcium, silicate, phosphorus, iron and inorganic carbon. These will end up in the ocean, where they contribute vitally to algae growth and provide a buffer against acidification. Everything is certainly connected, is it not?

## THE CARBON QUEEN EMILIANIA

*Emiliania huxley* is a beautiful name, but it also belongs to a beautiful creature. (I have tried to think of a more appropriate word than *creature*, but 'evolvature' does not have quite the same meaning. Maybe evolver or simply volver?). Emiliania, though, is not some beautiful mermaid and both the forename and surname have a masculine origin. The forename stems from the Italian-American researcher Cesare Emiliani, who is considered the founder of palaeo-oceanography. This not altogether intuitive concept conducts studies of past climates through analysis and dating of ocean sediments, which, owing to the presence of microfossils, have much to tell us about climate changes over hundreds of thousands of years. There is an eternal rain of dead algae and other particles falling into the ocean depths and becoming encased in the sediment for all eternity. Some have shells that can become silent witnesses to climatic conditions of ages past. *Emiliania huxley*'s predecessors are among these crucial witnesses to time.

The surname is also derived from an important historical figure. Thomas H. Huxley was not simply the defender of Darwin's theory, but an exceptional evolutionary biologist in his own right. Among his many accomplishments, he studied calcium flagellates as the microscopic building blocks of England's iconic limestone white cliffs of Dover. *Emiliania huxley*, known as Ehux to its friends, is a visually unique and aesthetically pleasing algae, although its most important characteristic is that, despite its small size (only five micrometres in diameter), it is a major player in regulating the earth's climate. Ehux is the ocean's carbon queen.

Carbon has acquired its own geological epoch, a fact due to the exceptional conditions that resulted in our today being able to excavate carbon buried over the course of 60 million years, thence returning it to the atmosphere at a furious pace.

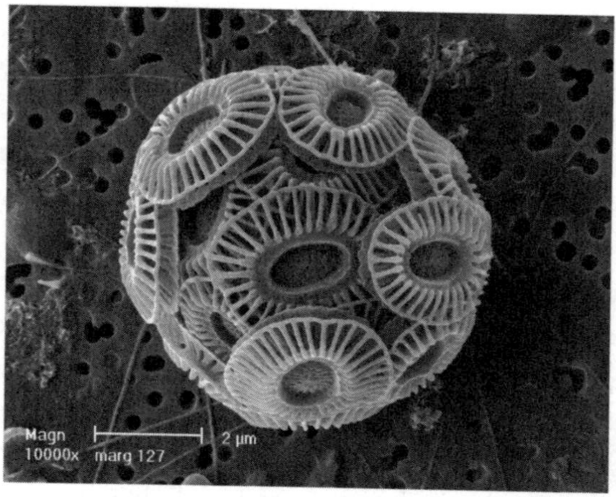


*Emiliania huxley* seen in all its beauty through an electron microscope. The calcified plates are one of the reasons why *Emiliania* plays such an important role in the marine carbon cycle, though they also make it vulnerable to marine acidification.

The carbon period, known as the Cretaceous, lasted from 145 to 65 million years ago and was characterized by the predecessors of today's Ehux, the mighty coccolithophores or calcium flagellates. The white cliffs of Dover, towering more than 100 m high, are largely built from the microscopic calcium carbonate shells belonging to calcium flagellates. We will not venture to estimate how many cells actually form the basis for Dover's cliffs, which, furthermore, were shaped in a warm, tropical ocean, but they are the cumulative result of tens of millions of years' production where $CO_2$ was transformed into calcium shells.[7]

Within its genus of calcium flagellates (there are around two hundred species in total), Ehux reigns supreme. It is also an evolutionary newcomer, making its first appearance in the ocean floor's fossil bed around 270,000 years ago and becoming dominant about 70,000 years ago. Purely from a time perspective, its history is not unlike the human one; and Ehux, like us, has enjoyed formidable success. Under favourable conditions – ample light together with access to phosphorus, nitrogen and

$CO_2$ – Ehux can form blooms that cover more than 100,000 square km and are visible from space as lighter ocean areas. Aside from binding $CO_2$, Ehux plays other roles in the earth's climate. Because it occurs in such large quantities, it can influence how the ocean reflects light over large regions, increasing albedo, the back radiation of light energy, and thereby leading to less solar energy being absorbed by the ocean. In addition, Ehux, along with a number of other algae types, produces dimethyl sulphide, an important component in the global sulphur cycle and critical to cloud formation – and, therefore, to the climate. Nothing proves better than Ehux that it is not size that matters.

Ehux's central role in the carbon cycle is not simply due to its $CO_2$ uptake, but to precisely the fact that it transforms a significant portion of $CO_2$ into calcium carbonate or calcite, much of which falls to the ocean floor where $CO_2$ is safely sequestered for the foreseeable future. Ehux, therefore, plays a central role in the marine carbon pump. At the same time, it is a species that potentially has the most to lose from ocean acidification, since calcium carbonate fares poorly at low pH levels, where the reaction is reversed and the calcium dissolves.

Why did Ehux and other calcium flagellates choose to equip themselves with decorative calcium plates? The ornamentation has no aesthetic basis, of course, but only a physico-chemical one. We might perhaps imagine that calcium plates provide some means of protection against hungry copepods and other zooplankton with algae on the menu, and it is possible that they do have a certain protective effect. Nonetheless, calcium plates are primarily an evolutionary answer to the chronic lack of nitrogen and phosphorus. Instead of cell walls that require such valuable resources, calcium flagellates form an inorganic cell wall that requires nothing more than carbon, which is usually in ample supply.

The ocean's most important grazers, copepods (2–3 mm in size, but still the ocean's equivalent of cows and wildebeests), gobble calcium flagellates with great abandon. The intestines of a copepod feasting on a spring Ehux bloom can be packed with up to 100,000 algae cells. After the valuable content beneath the shell is assimilated and transformed to copepod protein and copepod lipids, a prepackaged pellet of $CaCO_3$ is sent from the intestinal system to the ocean floor. Here we have another important part of the biological carbon pump.[8] If the ocean absorbs more $CO_2$ than it releases, thereby ensuring Keeling's Curve is not as steep as our emissions might suggest, Ehux takes part of the credit. It is always worth remembering that although the ocean's microscopic plants, despite their veritable number, comprise no more than 10 per cent of the earth's total plant biomass, still they are responsible for half the photosynthesis and carbon-binding that occurs.

As such, Norway's most important animals, for example, are not salmon or sheep, but rather the copepod *Calanus finmarchicus*.[9] This dominant zooplankton forms the link between algae and fish. In the central Norwegian Sea there is a calculated 200 million tons of *C. finmarchicus*, twenty times greater than the spring-spawning herring population. Some people believe *C. finmarchicus* is critical not just to fish, but to the ocean's supply of $CO_2$ and $O_2$. The argument runs that when these animals appeared an estimated 550 million years ago, the biological carbon pump kicked into action. Whereas before algae were broken down, surrendering much of their stored carbon back to $CO_2$, a massive transport of compact, carbon-rich pellets now arose, providing a significant export of carbon to the ocean depths. Since the breakdown of organic material in the upper levels so dramatically declined, oxygen use would also decrease. It sounds like a wild idea, millimetre-sized zooplankton having had such a significant impact on the earth's carbon and oxygen budget, but again we must remember that sheer quantity can offset a modest size.

Others argue that the most important group of 'organisms' in the ocean is actually viruses, which determine much of the fate of algae and bacteria – and, as such, notably impact carbon – though quotation marks are required here because technically viruses are not considered to be alive. Whales, on the other hand, hardly contribute to the ocean's carbon budget. Instead, all calculations suggest that the biological pump is primarily driven by the microscopic members of the ocean's food web.

Calcium flagellates, of course, are not alone in binding oceanic $CO_2$. There is also a rich assortment of other phytoplankton and larger algae (seaweed and kelp, for example) that convert $CO_2$ to organic carbon, which then enters the ocean's biological food web. Where this carbon will end is difficult to say. If you eat cod or a fat, carbon-rich mackerel for dinner, some of it will end up inside of you, where a minuscule amount will perhaps be stored, though most will be combusted and returned to the cycle. Of every ten C atoms bound as $CO_2$ by an algae, five will be returned to the atmosphere in cellular respiration. If that algae is eaten by a copepod, it might take up four of the remaining five atoms, perhaps burning one of those four before it, in turn, is eaten by a mackerel. Much of this $CO_2$ will never reach the ocean floor, but a steady rain of dead organisms and particles does actually make a difference and nothing does the job as effectively here as *Emiliania huxley* and *C. finmarchicus*.

An estimated quarter of the carbon bound by algae through photosynthesis makes its way to the depths, where it becomes food for bacteria and after that $CO_2$, though down here $CO_2$ enters the slow, deep ocean circulation. The portion that reaches the ocean floor to be more permanently stored in the bottom sediment is minimal, but it is still enough to make a difference in atmospheric $CO_2$ levels. Once again, Dover's white cliffs are a reminder that, given time, mountains can result.

The amount of dissolved, inorganic carbon in the ocean is around fifty times higher than in the atmosphere. On sufficiently long timescales, the ocean actually controls the atmosphere's $CO_2$ content, not the other way around. A continual $CO_2$

exchange occurs between the atmosphere and the ocean surface, an estimated 90 gigatons out and 92 gigatons in. In the upper, productive water layers, $CO_2$ is a starting point for Ehux and other primary producers, and its concentration is, therefore, limited by these, whereas in the depths $CO_2$ increases. This increase is partly due to the eternal rain of particles from the upper water layers, which is then partially dissolved, partially broken down by bacteria in the depths. The other source is a pump driven by the laws of physics. Some places, for example the Fram Strait between Greenland and Svalbard, are especially important for deep water formation. In these places, cold, salt-rich water sinks to the depths because cold, salty water has a high density (water increases in density the higher the salinity, and, at a given salinity, is heaviest at 4°c). Such places, therefore, draw $CO_2$ down to the depths, where it can remain for centuries while following the deep ocean currents. Sooner or later it will reach the surface again, but usually at more southerly latitudes. The Gulf Stream is a part of these current belts, where cold water in the depths makes its way south while warmer water comes to the upper layers in return.[10]

There are many reasons to worry that the ocean's critical carbon pumps and its ability to handle large quantities of $CO_2$ emissions are weakening. Since the addition of $CO_2$ happens many times faster than the slow addition of carbonate from weathering on land, what occurs is both an acidification and, eventually, a saturation of $CO_2$ uptake (more will be said on acidification later). Warmer water will provide a more stable 'cap' of warm surface water that, for many reasons, will prevent carbon transport to the depths, simultaneously reducing the supply of nutrients emerging from the depths. As such, algae's production and its carbon absorption ability will weaken. In short, we cannot count on the ocean to solve our problems. What about the forest?

## BOREALIS

Near my lake is a high ledge where I have hiked. On the west side is a precipice, and from there I can see the water in its context: a blue eye in a green sea. The green sea is part of the world's largest carbon storage unit on land: the boreal coniferous forest. It forms a broad, green belt stretching from Norway's west coast, through Sweden, hopping then over the Baltic Sea and broadening into a wider belt that stretches unbroken for many hundreds of miles through Finland and Russia, where it also covers parts of Kazakhstan, Mongolia and Japan, before continuing on the other side of the Bering Strait to cover large swathes of Alaska and Canada. The boreal coniferous forest comprises a third of the world's forests, but sequesters almost 60 per cent of the world's carbon.[11] In comparison, tropical forests account for 30 per cent of carbon sequestration in forests, and the rest of the world's forests account for no more than 10 per cent.

As such, it is no exaggeration to say that the boreal forest's inhalation and exhalation of $CO_2$, and particularly the difference between its inhalation and exhalation, is of vital significance to the earth's carbon budget. The forest, like the ocean, performs an invaluable service for us by providing a net $CO_2$ uptake. Much of this carbon is stored as cellulose and lignin, the two most dominant carbon polymers on earth. Together these two compounds account for well over half of the non-fossil organic carbon found on land.

A spruce tree normally survives about thirty years in a boreal forest before being felled. On its own, a spruce tree can reach well over a hundred years, a pine maybe four hundred. A venerable age, but sooner or later even the toughest mountain pine must surrender its carbon. Some trees can last a truly long time, for centuries even, and tree trunks have been found in mountain bogs in Norway that are several thousand years old. These are the remains of the forest that grew here in past warm periods.

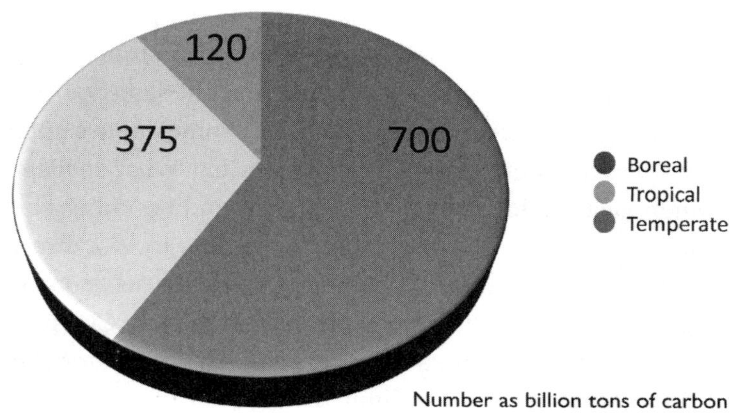


The distribution of carbon found in major types of forest.

The age rings of living trees, ancient wooden structures and such trees trapped in bogs can all be used to reconstruct the climate from a few hundred or even thousands of years ago. Poor years resulted in low growth and close rings, while favourable years had the opposite effect, but this does not provide a clear picture of past $CO_2$ levels. There are, however, other ways.

It is hard to imagine that putting a museum's dry, old herbarium plants beneath a microscope would show anything sensational. No doubt it happens rarely, but it *can* happen. Of course, we are talking about sensational in a scientific universe. In 1987, when Ian Woodward published his results concerning stomata in herbarium plants in the pre-eminent journal *Nature*, he knew biological circles would take note.[12] The results, however, were also relevant to climate research. The legacy left by Carl Linnaeus, for example, contains several thousand carefully pressed and well-preserved plants from the 1700s. Charles Darwin and many other naturalists followed suit, so for some plant species there is a timeline extending almost two hundred years. What did Woodward discover after months spent with dry plants in British museums? He found a significant reduction in stomata density the closer the plants got to the present, with current plants having around 40 per cent fewer stomata than their 150-year-old predecessors collected from the same location.

When I get low on $O_2$, for example after some intense uphill training on skis, I compensate by heavy panting. What do plants do when gasping for $CO_2$ – not after expending energy but because not enough $CO_2$ is present? According to Woodward, they increase their number of breathing holes, that is, stomata, and, likewise, they produce fewer when there is more $CO_2$ present. This is because more stomata means a higher water loss, and therefore it is useful to limit the number of openings to only what is necessary in order to conserve water. This theory was easily confirmed by cultivating plants at different $CO_2$ concentrations. Indeed, the number of stomata varied in keeping with the $CO_2$ present. As such, herbaria collections could tell us about past $CO_2$ concentrations – and not just for two hundred years or so. Further research eventually showed that fossilized plants, many of them preserved down to the smallest detail, even to the smallest stomata, also contained a climate archive, or $CO_2$ archive, that could offer a glimpse of $CO_2$ levels millions of years ago.

FROM MY high vantage point, I gaze with thanks on the massive green lung surrounding me. I experience the forest's presence as invigorating. The green carpet stretches as far as the eye can see over ridges, dales and into blue distance. Now in the sunlight an incomprehensible number of $CO_2$ molecules are passing through an incomprehensible number of stomata into an incomprehensible number of coniferous needles. The sugar formed by photosynthesis will be transported around, the bulk of it hooked into polymers that enter cell walls and build cellulose and lignin. This process takes energy and plants also burn sugar in order to create the energy. A significant portion of the carbon bound by photosynthesis, therefore, also returns to the atmosphere. The tree will also direct much of the sugar down to the roots where it will fire cellular respiration before being set free as $CO_2$.

This subterranean $CO_2$ has different and more important functions than the $CO_2$ exiting leaves, needles and stomata. Tree roots reach down into the slow, weathering-based carbon cycle because the $CO_2$ produced below ground does what $CO_2$ usually does when it meets water: it forms carbonic acid. Carbonic acid in contact with earth, stone and bedrock helps to release from the ground minerals needed by the tree, such as phosphorus, iron and silicate. This forms a kind of biological $CO_2$ pump, because such weathering processes bind much of the $CO_2$ retrieved by the tree's other parts from the atmosphere.

Coniferous forests, which circle the north of the planet like a belt, keep most of their large and long-term carbon banks below ground, in roots and in soil. Three-quarters of all the carbon in the boreal coniferous forest is located beneath the turf and in it. This is the result of careful bank deposits over many millennia and here the carbon is largely kept secure. There is also a kind of interest on this deposit, because more soil means better forest growth, which results in more soil. The total quantity of carbon that is sequestered (gross production) minus that which is set apart as $CO_2$, or leaked out as sugar, is what remains to the tree. That forms the net production and is what counts for forest users as well as for the carbon budget. On the other hand, the $CO_2$ molecules that take a quick tour through the plant only to leave it again are part of a zero-sum game in carbon terms.

A given tree can be involved in many types of cycle on many different timescales. Most of the $CO_2$ the tree absorbs returns to the atmosphere after a quick trip through the plant's cells on a timescale of less than a day. On a timescale ranging from a day to a year, anything that gnaws or sucks on the tree, such as bark beetles, aphids and other assorted grazers, will transform some of the wood into bodily tissues and their cellular respiration will return $CO_2$ back to where it came from. The tree's annual cycle means that needles and leaves also fall, decompose and return $CO_2$. If the tree is felled, twigs and bark will release their carbon relatively quickly following a massive assault from insects, fungi and bacteria. The wood itself might have a lifespan of a few years as paper or decades as building material.

In the long term, what really counts when it comes to the boreal forest's carbon sequestration efforts is the carbon found below ground. In the forest's subterranean depths, there is also an ecosystem of fungi and bacteria, but cellulose down here is poor fare, so the carbon can rest in peace for thousands of years beneath the tree crowns' shadow.

Of course, carbon sequestration in a boreal forest also varies enormously. Over the ridge above my lake the mountain clearly rises today, 8,000 years after the ice abandoned it. Here, where the firs cling fast, it is perhaps 10 cm to the bedrock and the roots creep along the face to find a foothold, hunting for cracks or uneven places into which they can burrow. A couple of pines have succumbed to last year's autumn storms right below the cliff where I sit, illustrating just how difficult life is here on the ridge. The roots expose clear bedrock beneath a sparse layer of soil, the bedrock scoured by the glacier when it withdrew. After 8,000 years, this thin layer of soil is all that carefully labouring plant generations have succeeded in depositing in the bank.

Right below me is a valley basin. If you stick a spade into the earth there, you find black forest soil and it takes real work to reach the bedrock. Along a bog pool next to the lake, there is a 'bottomless' layer of black swamp soil built up beneath the peat. Here a crowbar could be sunk into the ground without ever hitting bedrock. How can 8,000 years of history leave such different carbon stores in the very same region? Much of the answer lies with water. The terrain sticking up in the air stays dry, there is never a lack of oxygen, and tree litter, branches and trunks will fully break down into $CO_2$. In contrast, water collects in the basin, the decomposition process is slower, a soil layer builds

Not much carbon has been stored in the shallow soils in elevated areas of the boreal forest since the last glacial period. The trees are clinging to a thin layer of soil on top of the bedrock that is exposed to windfalls. In the lowland areas nearby, however, there are bogs and wetter areas with metres of black, carbon-rich soil.

up that provides a basis for better plant growth, and over time the slightly deeper layers will become oxygen deprived, slowing decomposition even more. The ice probably also contributed to the fact that the peaks were scoured clean of debris, whereas the basins retained glacial sediment and finely distributed material.

In bogs the combination of oxygen deprivation and a low pH due to humic acids help halt decomposition, something to which bog bodies can attest. Grauballe Man, who was uncovered in 1952 during peat excavation in a bog near Silkeborg, Denmark, looked as if he had stepped into the swamp a year ago.[13] Dating using 14C, however, showed that he was dumped into the mire, into what probably was still a bog pond, about 300 BC. An autopsy revealed a brutal death caused by a slit throat and a crushed skull. The man's last meal was shown to be barley porridge, wheat, herbs and meat, probably pork: this indicates that the carbon preservation was reasonably complete. Bog bodies in equally well-preserved condition are less commonly found in Norway, but countless discoveries in the highland bogs of roots that may be dated as several millennia old illustrate not just the preservative effect of bog ponds on carbon, but also that the Norwegian mountains had an entirely different forest boundary, with the southern mountain plateaus being covered by forest until the Bronze Age. We will probably see a forest-covered plateau again in two or three hundred years. In that case, it will contribute to a significant increase of carbon sequestration in the forest, which brings us to a lively, not to say heated, subject of discussion. How can we use forest for carbon sequestration?

THE PROSPECT of forest serving as a climate mitigation option is being discussed in most countries with large forest-covered regions, such as Canada and, perhaps most especially, Norway. There can be multiple reasons for this, some of which are obvious. Norway has a real need to brush up its carbon accounts (something it shares, for good or for bad, with many other countries), although its energy sector is already largely based on carbon-free water power. On the other hand, the country is a major actor in the fossil fuel industry, there is the prospect of expanding and intensifying the logging industry, and the beleaguered pulp and paper industry is searching for new markets beyond these materials. In principle, every petrochemical product can be retrieved from the forest; the timber industry often claims it can deliver everything the oil industry delivers, and if trees can provide fuel in addition to managing $CO_2$ emissions through photosynthesis, it is difficult to get any cleaner and better. There is no doubt that, climatically speaking, energy from trees is better than energy from coal, oil or gas, so long as we assume that the $CO_2$ released from forest-based bioethanol, for example, will again be absorbed by the forest. The ideal would be a closed loop that made no net contribution to the atmosphere (trees obviously will not distinguish between 'clean' $CO_2$ from bioethanol and 'dirty' $CO_2$ from oil).

This discussion largely centres around which forest type is best equipped to help us manage our carbon balance: an intensively managed plantation forest with young, fast-growing trees, or an old, slow-growing forest where nature takes its course and fungi are allowed to feast on rotten trees when they reach the end of the line.

The answer, for many reasons, is not as obvious as one might think. Fast-growing trees will certainly bind plenty of carbon, but it takes a few years before the trees reach a size where they truly can make a difference – and here come the logging machines. The way is certainly cleared for the next generation, but the question then becomes, what happens to all the removed tree material in the meantime. The majority will end up as pulp and it will not be long before paper returns its borrowed carbon to the atmosphere. The process takes a little longer for whatever becomes building material, but even a house does not last forever, so this is no long-term savings plan either. If all Norwegian timber ended up in stave churches, like those built from the twelfth to the fourteenth centuries, that would be another matter, aside from the fact that fast-growing spruce would not be a good bet in a stave church that could survive for a thousand years. Still, purely in terms of volume, many northern forests are already experiencing fast growth. Norwegian forests, for example, are currently handling about 25–30 million tons of $CO_2$ annually, close to half of Norway's mainland $CO_2$ emissions.[14] Because more forest will necessarily bind more $CO_2$, the potential here is undoubtedly immense. Yet once again the pressing question here is time. There are limits to how far forests can expand and, over the course of a relatively short time horizon, the potential will be exhausted.

Even though the intensively managed forest has its limits, an old forest where trees are allowed to rot where they stand will probably have much less to offer. This, however, is not necessarily true. Old forests (or, more particularly, mixed forests with different types and ages of trees) bind a lot of carbon and they seem to be more effective, especially when it comes to depositing carbon into the subterranean bank, which, after all, is where the majority of forest carbon is located. In fact, the key here seems to lie more with the fungi than the trees, although there exists a symbiotic relationship between the mycorrhizal fungi found throughout forest soil and tree roots. As it turns out, fungi are responsible for the majority of carbon sequestration taking place in forest soil, and because these fungi thrive better in older, mixed forests, the evidence indicates that they perform better than in young forests, given equally large carbon deposits and a more long-term savings account.[15]

Truly large forest fires are a rarity in Norway, in part because actual drought is not so common, any fires are quickly extinguished and there is only a modest area to manage. In Canada, on the other hand, the boreal forest is huge and the same goes for forest fires. They are also extremely common. These fires have, in fact, been shown to be critical to the forest's carbon sequestration. As such, it is perhaps not surprising that there is a sizeable debate regarding whether the fires should be extinguished or whether they should be considered part of the forest's natural dynamic and so allowed to burn. Perhaps more surprising is the fact that, whereas Canada has protected enormous swathes of boreal forests in honour of carbon sequestration and biological diversity, the world's most extensive tar sand extraction is under way there, not only destroying forest and diversity, but employing one of the least energy-effective and most polluting methods by which to extract oil. The paradox is located, of course, in the autonomy and perspective on reality adopted by the individual Canadian provinces. It seems just as contradictory that Norway devotes billions of kroner to rainforest preservation, while, via the oil fund, investing in companies that demolish that same forest. In all fairness, however, it must be said that there was some improvement once Norway recognized the paradox.

The carbon budget also comprises several factors, since there are hardly limits to the products that cellulose and lignin polymers can give us. For example, it is fascinating that bland lignin can be split to form a tasty variety within the numerous CHO-family, something with the formal name 4-hydroxy-3-methoxybenzaldehyde ($C_8H_8O_3$), which consists of the standard six carbon ring with an O, a hydroxyl (-OH) and an aldehyde group (-$OCH_3$) linked on. The result is better known as vanillin, the basis for vanilla. It seems tree trunks can be found in vanilla ice cream and custard as well. Even more fascinating is the fact that lignin also forms the basis for slightly less delectable products, such as asphalt, concrete and paint. Cellulose can also lend its polymers to everything from textiles to bioethanol. The day we undergo a closed carbon cycle from forest to fuel to $CO_2$ and back to forest is the day the forest industry can truly boast about with pride.

The balance sheet also contains components that cannot immediately be translated into carbon. The forest provides workplaces and tree products that we need, but it is also home to a large biological diversity that thrives poorly in monotonous plantation forests. One of the species sometimes found in forests is people, who seem also to seek out older, diverse and more species-rich forests. That being said, there is no easy answer to which is the best type of forest, simply because the answer depends on your viewpoint. Such is the way with all value questions, but when it comes to the carbon budget, and especially given a slightly longer time horizon, much indicates that the old forest is significantly undervalued.

Many of us assume that the boreal forest, in contrast to the rainforest, will increase in scope as the warmth creeps north, thawing tundra and taiga and preparing the way for a northward forest expansion. And that is indeed what is happening. In any case, it has become indisputably greener as bushes and shrubs head north across the tundra and over the mountains. The status of the boreal forest, however, is also uncertain. Satellite data indicates that northern forests are losing ground owing to logging and development. At this point, the reduction is marginal compared to what is happening farther south.

## PARADISE LOST

The carbon balance looks entirely different in tropical forests. Life there is largely lived hand-to-mouth, without the ability to deposit any noticeable carbon in the bank. Even though the world's rainforests still cover a substantially larger area than the enormous boreal coniferous forest belt, they contain only half as much carbon. The explanation for this is the simple fact that almost all the carbon in a tropical forest is found above ground level in the trees themselves. There is some stored in the roots, of course, but anyone who sticks a spade in the earth in a tropical forest will usually find that they strike hard, reddish, mineral soil, completely different from the black soil found in northern forests. Any leaves and trunks that fall to the ground will almost immediately return to the atmosphere as $CO_2$ after insects, fungi and bacteria have finished with their feast, whereas the fate of fir needles and fir trunks is entirely different. In the northern forests, of course, the collective effort of all the various actors behind the decay phenomenon will play their part and much carbon will end up in the atmosphere. A significant portion, however, will also be trapped and transformed to humus layers and soil. The classic explanation for this difference between tropical forests and northern forests is that cold prevents the decomposition process from keeping pace with net carbon sequestration, whereas everything in the warm, tropical moisture decays quickly. To some extent this is correct, but the real difference is actually due to the fungi found in the soil.

We have fully recognized only recently that fungi are actually the major conductors behind the forest floor's carbon cycle.[16] Tropical forests are home to a fungi flora that handles with reasonable ease the complex polymers found in cellulose and lignin. After these do their part, other fungi and bacteria can enter and the party can truly begin. In northern forests, on the other hand, brown-rot fungi prevail, a species largely tied to coniferous forests, as well as oak, beech and birch. Brown-rot fungi leave most cellulose and lignin untouched, and this fact, combined with the low temperatures, contributes to black soil full of carbon. A rainforest in equilibrium, in contrast, does not contribute either to the plus or the minus side of the atmosphere's $CO_2$ content (nor to its $O_2$ content). This means that the metaphor of the rainforest as the earth's lungs is not especially apt, since a lung obviously takes in $O_2$ and exhales $CO_2$. The moment it is felled or burned, meanwhile, a carbon transfer will occur from wood to air, gradually with cutting, spontaneously with burning, and the focus on the rainforest is entirely legitimate. Ecosystems that have taken tens of millions of years to build up with their unique species diversity are, in the course of a historic gamma ray burst, in the process of being lost forever. More than carbon and the climate is at stake here, but since carbon is our theme, we will leave colourful macaws, tapirs, primates and a thousand other species alone for now.

The loss of these rainforests is happening at an unimaginable rate. For nearly all of human history, rainforests have covered between 10 and 15 per cent of the earth's surface.[17] As late as 1970 most were still intact. The world's population at the beginning of that year was officially 3.692 billion; today we are double that, whereas rainforests have been halved. Between 2000 and 2012 alone, 130,000 square km of rainforest disappeared each year, an area about the size of Greece. In the world's largest rainforest, the Amazon, 216,000 square km disappeared between 2000 and 2010, which is about the size of Great Britain. And so we continue to clear unbelievably large areas in an unbelievably short time. If this tempo continues, and it seems that it will, only fragments will be left – and fragments of rainforest will soon cease to be rainforest.

Brazil, the world's largest rainforest nation, actually takes its name from the Portuguese word *brasa* or 'ember'. Charcoal is an old Brazilian export, though today it is only a marginal energy carrier in relation to coal. Still, it is not so marginal that it does not pose a significant threat to the world's rainforests, thereby also affecting the global carbon cycle. In total, around 1.35 million tons of charcoal are used annually, which still pales in comparison to the annual consumption of coal, which equals about 7 billion tons. This also places significant pressure on the Amazon, exerted primarily by the local iron industry: the largest ironworks in the Amazon requires at least 12 million cubic metres of timber each year, equivalent to clearing 200,000 hectares of forest annually.

For now the rainforest appears to absorb more $CO_2$ than it emits as the result of a rather paradoxical $CO_2$ fertilization effect. Although normally speaking nutrients like nitrogen and phosphorus limit plant growth, more $CO_2$ can increase growth for a time. That means more $CO_2$ passing through plants' stomata, which kicks photosynthesis into high gear. More sugar is thereby produced and more polymers are stored. The result is more plant volume, but this occurs without an equal amount of other elements taken up, resulting in poorer plant quality for grazers, and the decay of plant materials happens at a slower rate. This situation is also necessarily temporary, but for the present the rainforest probably also contributes to the fact that the Keeling Curve is not steeper.

A recent study of the Amazon has calculated an uptake of 2.2 billion tons of carbon and an emission of 1.7 billion tons, although creating a carbon budget for the Amazon is not as trivial as for a Norwegian forest lake – which itself should not be considered trivial. The Amazon's carbon balance also contains significant differences between dry and wet years, and in a dry year there can be a net exhalation of up to 500 million tons of carbon back into the atmosphere. The Congo rainforest has declined in greenness as a result of drought and the same has happened in the Amazon.[18] Some researchers fear that the Amazon is caught in a downward spiral due to logging, burning and fragmentation, all combined with higher temperatures and increased drought stress. Satellite images also show that since 2000 the forest has decreased not only in scope, but in greenness. The Amazon seems to be experiencing breathing problems and might end up sweating out more $CO_2$ than it absorbs. One nightmare scenario would be the gradual desiccation of the entire Amazon toward a pampas-like landscape. Such a shift would certainly not go unnoticed in the global carbon cycle.

There is carbon on all sides of the equation here. On an annual basis, forest loss (wood transformed to $CO_2$) contributes to 10 to 15 per cent of human $CO_2$ emissions. Why is this? No one actively dislikes the rainforest: we are just increasing our numbers and our consumption, and are consuming in the wrong way. Half the world's loss of rainforest can be attributed to an increasing demand for four products: palm oil, soy, timber/ paper and steak. So what exactly is steak's true carbon footprint? We can ask this question even of steak produced locally. For example, although cows in Norway do not graze on pasture that was once a rainforest, livestock production is still largely based on imported soy, a crop often grown in cleared rainforest areas, so the net effect is still the same. For its part, Norway is actually the world's third largest soy importer per capita, which means more than 530,000 tons, chiefly from Brazil, and chiefly for concentrates and fish feed. The space required to produce this amount of soy equals 17 per cent of Norway's agricultural area. As a result, all Norwegians contribute to deforestation. The same would be true of any other country that imports its soy from Brazil. On top of this, we also import a generous portion of Brazilian beef. As we see, the carbon cycle has many labyrinths and, with a globalized flow of goods and services, accountants must remain hypervigilant in order to keep the books correctly. For example, should $CO_2$ emissions that result from deforestation due to production of palm oil, soy and beef be billed solely to Brazil, Indonesia and the Congo – or should countries like Norway also be included?

Some rainforest areas also have a carbon bank sequestered below. The wetter it is, the greater the chance for oxygen-free conditions, which means less organic carbon being oxidized to $CO_2$. The Amazon's wettest regions harbour significant carbon stores, but nothing competes with Indonesia's peatlands, where carbon has continuously been deposited in the bank for millions of years. These swamp forests are home to orang-utans and thousands of other species found nowhere else on earth, and the forests stand on top of enormous carbon stores. In some places the peat layer is up to 20 m thick and in many places over 10 m. In fact, Indonesia's peatlands contain as much carbon as the whole Amazon rainforest. Whenever the forests are cleared, the swamps drained and, in the worst case scenario, burned (peatlands can burn for weeks or months), millions of years of sequestered carbon is returned to the atmosphere in the blink of an eye.

Swamps comprise an undervalued ecosystem that has been largely viewed as having value only to the extent it could be drained, excavated and farmed, and eventually harvested for peat fuel. Historically the world's wetlands and bogs have been treated unfairly, having been transformed on a large scale to more agricultural-friendly areas, simply because here their reputation was good: once the water was gone, things grew really well there.

Whittlesey Mere is a fertile area near Peterborough, north of Cambridge.[19] Until 1850 it was a large, green lake, the largest in southern England, although much reduced in size by seventeenth-century land reclamation schemes throughout the Fenland area. The building of a pumping station equipped with the latest Appold centrifugal pump made it possible to convert most of the lake to fertile farmland. It was then that a local landowner, William Wells, had an idea, something like a low-tech variant of Keeling's $CO_2$ measurement. Wells observed that the land sank almost immediately after being drained. To measure this, he had a massive iron pole driven into the ground and anchored securely. At first it was just the top of the pole that stuck above the grass, but over the next decade it already stood 1.8 m above ground level. The sinking slowed somewhat after that, but now the top of the pole is 5 m above ground.

Whereas Keeling documented a steady carbon accumulation in the atmosphere, Wells's simple measuring stick documented carbon as it disappeared from the ground. What happened first was desiccation, which then caused the wetlands' enormous carbon stores and old lake sediment to oxidize, to transform into $CO_2$ and vanish into the air. It took maybe 10,000 years to build up all those metres of carbon-rich swamp and soil through photosynthesis, but only 150 years to reconvert it into $CO_2$. The principle is the same for our returning coal, oil and gas to the atmosphere: the carbon that photosynthesis carefully deposited over thousands of millions of years into the ground is converting back to $CO_2$ in a historical blink of an eye. Exactly how much $CO_2$ has returned to the atmosphere via the oxidation of Whittlesey Mere is difficult to say, but estimates surrounding the annual degassing of $CO_2$ from drained English wetlands show a figure that can amount to 6 million tons of $CO_2$ per year. That equates to 15 per cent of annual Norwegian $CO_2$ emissions, a number that can range from 5 to 7 million tons. The point here is that soil and wetlands amount to massive carbon sinks – and therefore to massive carbon sources when they are converted to $CO_2$.

The full extent of the ecosystem services that swamplands perform for free in the form of carbon sequestration is still unknown. Destroying swampland always releases significant quantities of $CO_2$. It has been estimated that removing peat from a modest area of peat alone will result in $CO_2$ emissions equal to about 150,000 cars after a year has passed.[20] Of course, the swamp will not emit methane once it is gone, but it is obvious the $CO_2$ balance sheet will suffer a negative. Every year 5,000 acres (700 football pitches) of swampland are planted, excavated, dug up and developed – and that is in Norway alone. This represents a huge problem not just for the climate, but for the unique ecosystems that swamps and wetlands represent.

Ecosystems, whether they are lakes, oceans, forests or swamps, have a cycle that is intelligible, at least in principle and qualitatively. Quantitatively the matter is not quite so simple, but there is reason to believe we are working within the correct scales when it comes to stores and fluctuations. The carbon that plays a role in these cycles, which are largely biological, varies on a timescale of days (for obvious reasons, $CO_2$ concentrations in a forest are lower during the day than at night), years (winter–summer), decades (the forest's lifespan) and centuries (old forests, swamps, deep ocean). Here we are talking about processes and timescales that strike us as comprehensible and relevant. Still, these cycles are connected to large cycles that take place on completely different time scales. Zooming in, we can regard the Amazon's carbon budget, for example, as the sum of every tree's absorption and emission, which again is the sum of every leaf's absorption and emission, which again is the result of each individual leaf's numerous stomata and chloroplasts coupled with myriads of insects, bacteria and so on. Research is being conducted on each of the scales, all of which represent pieces in a carbon puzzle we will never finish. In order to operate with a manageable number of pieces, we are required to zoom out and wield larger pieces. The image becomes coarser, with fewer pixels, but it also provides an overview. Furthermore, it gives us some desired numbers that at least allow us to say whether 10 tons or 10 gigatons is a lot or a little.

## THE LONG CARBON CYCLE

Carbon can perhaps be termed life's central building block, but alone it has little value – that is, if we are thinking in terms of life and not diamonds. Life requires a combination of the right ingredients in the right place at the right time. Indeed, life presupposes such an absurd complexity of intricate molecules in intricate connection that we quickly meet a reduction problem. Remove any one of these components and life grinds to a halt: life simply requires a great complexity of things to be simultaneously present in order to function. And what about the magic spark that set it all in motion? Humanity has a long tradition of explaining the inexplicable with the help of mysterious and incorporeal powers, which amounts, strictly speaking, to replacing the incomprehensible with something even more incomprehensible. When it comes to life's origin, we can well be humble enough to admit that we may never know the ultimate explanation. We know enough, however, to say how the path is laid, stone by stone, from the rather random, albeit critical, transition point from non-life to life – to complex individuals with billions of cells.

The first life forms had 'only' two conditions that must be met: there had to be a kind of stable unit, preferably a surface or membrane, and a means of self-replication. Replication is carried out by nucleic acids (DNA and RNA, though it is hard to say which came first). Viruses are at the transition point between non-life and life. They have a simple protein membrane that surrounds a minimal, but sufficient quantity of DNA or RNA. That does not mean that viruses are the origin of life – they depend, after all, on a host – but they demonstrate how subtle the transition point can be between non-living and living. From the very beginning, proteins could well have formed the membrane basis as well as acting as building blocks for most else that was required. Of course, proteins themselves are complex, three-dimensional and often large molecules, but they are constructed according to a simple principle. They are chains of amino acids linked together and there are only twenty amino acids in total. The composition of these acids, in turn, is governed by the order of the four-letter genetic alphabet. Where do the building blocks in proteins and nucleic acid (DNA and RNA) come from?

Here we encounter another name on the list of sharp minds to deliver critical insights surrounding carbon, Harold Urey, who began as a zoologist before switching to chemistry.[21] Urey studied under Niels Bohr and became associate professor of Chemistry at Columbia University in 1929. In 1934 he won the Nobel Prize for his discovery of deuterium, better known as heavy water, and as Director of War Research at Columbia he was also central to the development of the atomic bomb. Perhaps prompted by awareness of the atom bomb's destructive force, after the war

Urey became increasingly preoccupied with life, particularly with life's origin.

In 1952 he helped his student Stanley Miller set up an experiment to explore the extent to which complex molecules could develop in conditions presumed to be present in earth's early history 4 billion years ago. They mixed water, methane, ammonium and hydrogen, and then sealed the mixture into a sterile five-litre glass bottle further connected to a half-litre bottle half full of water. The smaller bottle was heated to boiling point and water vapour poured over into the larger bottle, which held the 'prebiotic' ingredients, that is, elements assumed to be present on the young planet prior to life's appearance. The mixture was subjected to repeated electrical shocks before the whole thing was cooled and the procedure was repeated. After only a day, the solution acquired a pinkish tinge, and after a week the experiment was finished and the contents analysed. Amazingly enough, they found that no fewer than eleven amino acids had been formed during just that one attempt, making it one of history's most amazing experiments.

One can, of course, rightly argue that this experiment proved nothing with regards to life's origin, aside from the non-trivial point that a reduced carbon compound in its simplest form, $CH_4$, that is, a reduced nitrogen compound ($NH_3$), water and hydrogen, could make enough of a spark to produce amino acids. After Miller died in 2007, the bottles were again retrieved. Fixed by mercuric chloride, they had been stored and sealed for 55 years. When they were re-analysed using more advanced methods than Urey and Miller had at their disposal, it turned out that all the rest of the twenty amino acids were present.

The actual process by which life originated is something we will never know, but it is an undeniable fact that life appeared – and the rest is history. Sufficiently stable units with self-replicating capability were required to start evolution, and here we are today, together with the millions of other descendants that have played their part to produce liveable conditions on earth through their contributions to the carbon cycle. Paradoxically, it must have been the hostile conditions 4 billion years ago that paved the way for life to form.

If Urey has earned a place in the history surrounding carbon, it is certainly not for his role regarding the atom bomb, nor is it solely due to his contributions regarding early atmospheric composition. It is not even for his critical insights into how life might have arisen, even though that was basically the origin of the life-driven part of the carbon cycle. Instead, Urey's central contribution in this context is to understanding the geochemical carbon cycle. As we shall see, this cycle is directly connected to the biological one, where biology can be regarded as a catalyst that helps gear up the comprehensive weathering process.

Among the many significant questions Urey pondered was: why does the earth's atmosphere contain such surprisingly little $CO_2$? On Mars and Venus $CO_2$ is much more dominant in the atmosphere. Mars has around thirty times more $CO_2$, whereas Venus has 300,000 times more (on such nearby planets, gas composition can be remotely measured with great precision). At the same time the earth has enormous amounts of carbon sequestered in rocks, sediments and coal. Urey concluded that there had to be some mechanism that perpetually removed $CO_2$ from the atmosphere. In his book *The Planets* (published in 1952, the same year as his ground-breaking experiment with Miller), he posed the following explanation: $CO_2 + CaSiO_3 \rightarrow CaCO_3 + SiO_2$, otherwise known as the Urey Equation.

A key process in the earth's long carbon cycle occurs when $CO_2$ is bound during weathering in reaction with $CaSiO_3$ (or equivalent compounds), thereby producing $Ca^{2+}$ and $HCO_3^-$, which rivers then carry from land to sea.

Here some will enter the biological cycle, though most will precipitate as $CaCO_3$, sink to the bottom and in that way be taken out of circulation. The same fate also meets the carbon

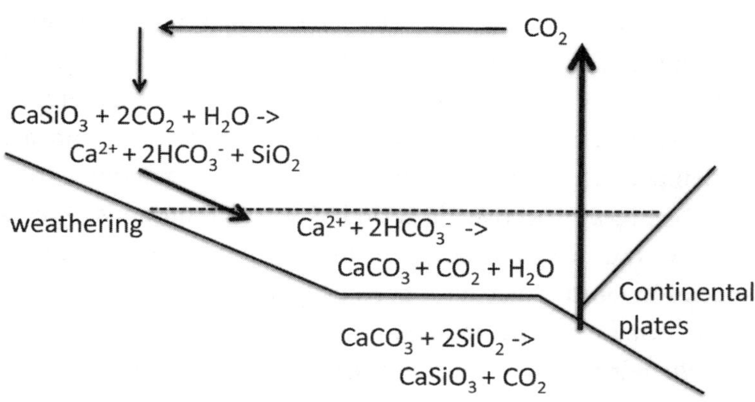


A simplified version of the slow geochemical carbon cycle, with the Urey reaction as a key component. Atmospheric $CO_2$ is bound by the weathering process, which also releases key elements for primary production. In the oceans, precipitation of inorganic carbon may occur. This carbon could be locked into the rocks for eternity, but some may enter geologically active areas where the rock melts and volcanic activity emits $CO_2$ back to the atmosphere.

that ends up in the calcium shells of algae or animals and which joins the eternal rain of organic particles to the floor. Here we are again reminded of Dover's cliffs. They are remarkable not for their immense quantities of chalk, but for the fact that these have been made visible in such a spectacular way.

Much ocean-sediment carbon will be locked away in mountains for all time, but as mountains are exposed to water and erosion, they experience perpetual wear and a stream of minuscule particles is constantly being fed into rivers, thereby contributing to mountains again being converted to ocean sediment over endlessly long periods of time. Mountains containing organic carbon ('petrogenic organic carbon') contribute a large portion of these particles,[22] much of the carbon of which will be oxidized back to $CO_2$ on meeting with air and water. Some of the carbon ends up in geologically active areas where the earth's plates either collide or separate and where high temperatures bake the mountain, before volcanoes again exhale $CO_2$ from the earth's depths. The earth surface's movement, plate tectonics, is a central factor in the slow-working thermostat that contributes to the fact that those claiming 'it will all take care of itself', whether the topic be global warming or ocean acidification, are probably right. That demands, however, that we adopt a truly far-reaching perspective – a period of several thousands of years – and we must realize that, in the meantime, most of the globe's species inventory may have been caught in the undertow, and, in the worst case, that includes ourselves.

When it comes to the *lithosphere*, what we can call 'the bedrock', carbon does not play a dominant role compared to other basic elements. In fact, it comes in at a humble fifteenth behind other elements, far below the top five, which are oxygen, silicon, aluminium, iron and calcium. The earth's bedrock actually contains only 0.03 per cent carbon, practically the same amount found in the atmosphere. The universe as a whole has little room for anything other than hydrogen (91 per cent) and helium (9 per cent), with carbon comprising only 0.02 per cent.[23] Here it is worth remarking that dark matter, which makes up an estimated 63 per cent of what we consider to be concrete matter, is not figured into the equation and neither is mystical dark energy, of which there is three times more than dark matter. From this perspective, carbon seems as insubstantial as we might feel at night under a starry sky. If we turn our focus to the biosphere, however, that is, to what we can generally call the sum of the ecosystems, carbon recovers a key position – at any rate, if we discount water.

Even if percentage-wise there is not much carbon in the bedrock, still it has absolute meaning within the carbon cycle. Carbon buried deep in the bedrock has only a slight chance of ever seeing the light of day, unless it randomly hitches a ride during a volcanic eruption. From a geological time perspective, one in which mountain chains arise and erode, a portion of the bedrock's trapped carbon will get the chance to participate in the cycle in the meeting with $CO_2$ and water and through Urey's weathering reaction. This sets in motion enough carbon to play a decisive role as the earth's long-term thermostat by binding $CO_2$ and simultaneously releasing life-giving minerals into the biosphere through this same weathering process.

If weathering's extremely slow carbon capture could be speeded up, many of our problems would be solved. So, could it happen? Historically two significant processes have particularly geared up weathering. Mountain chain formation exposes large quantities of 'fresh' bedrock to weather, wind and $CO_2$, and steep mountainsides will more easily export weathering products such as silicon and phosphorus, which then fertilize the ocean.[24] This will increase the binding of atmospheric $CO_2$ owing to increased weathering and increased oceanic algae production. The formation of the Himalayas apparently contributed to a global cooling on the basis of this principle, though not too quickly, of course. Whereas the Indian and Eurasian plates seriously collided 50 million years ago, much subsequent mountain formation took 'only' 10 million years. Geologists are impressed at the haste with which this occurred, but since time is short, we have to look for something else that can do the job on a different timescale.

Olivine is a rock from the earth's depths that proves to be not only a source of methane on its way up to the surface, but a source of $CO_2$. On their way up to the surface, gases are boiled out of the rock, so to speak, and the rock reaches the surface with a craving for $CO_2$. Like calcium silicate rocks, for example, olivine can absorb carbon. It will function as an infinitely slow sponge that sucks up $CO_2$ and, if you can help olivine better its absorption capacity, there is enough olivine available to handle current $CO_2$ emissions over a period of 3,000 years, given that we fill to capacity all the olivine down to 5 km below ground. The problem is that olivine, like any dry sponge, expands as it absorbs $CO_2$, ultimately increasing its volume by 80 per cent and so raising the terrain a few hundred metres.[25] No quick fix here either.

## THE ANTHROPOCENE

In the 1980s an environmental threat reared its head and overshadowed the 'climate crisis', which was not yet fully recognized outside the scientific community. This threat was the 'ozone hole', the rapid breakdown of the protective layer of ozone ($O_3$) in the stratosphere about 30 km above the earth's surface, largely due to our emission of chlorofluorocarbons that 'consumed' $O_3$. The man who described this process, pointing out the risk and ensuring that ozone-hostile materials were phased out, was Paul Crutzen. He received a well-earned Nobel Prize for his efforts. For Crutzen, the ozone story was just one of many examples of the way in which humanity, through sheer numbers, consumption and an impressive, but rather short-sighted, ingenuity was capable of fundamentally transforming the planet. In short, we were entering a new geological epoch that could best be characterized by the presence of one species: *Homo sapiens*.

Crutzen conceived the idea during a conference when the session leader repeatedly referred to the Holocene, our current geological epoch, or rather, our *previous* geological epoch, if we accept that we have now entered a new one. In that case, the Holocene indeed had a short career, since it began at the end of the last ice age 12,000 years ago. 'Hold on,' Crutzen burst out. 'We are no longer in the Holocene. We are in the Anthropocene.' Crutzen set out six points for discussion in his 2002 article 'Geology of Mankind': human activity has reshaped between a third and a half of the earth's land surface; most of the earth's large rivers have been forced to change course; the fertilizer industry now binds more atmospheric nitrogen than all the world's land-based ecosystems combined; fisheries commandeer about a third of all coastal waters' primary production; we now use more than half the world's easily accessible freshwater resources; and, last but not least, we are dramatically changing the globe's gas composition.[26] To this we might add a long list of other global impacts, such as species extinction and the fact that during the last forty years we have reduced the world's animal populations by 50 per cent on average. The Anthropocene has acquired a half-official status, complete with its own journal containing scientific articles that cover different facets of human impact, and so the concept is established.[27] Did the Anthropocene begin with agriculture 10,000 years ago, the extinction of megafauna before that or with the documented increase of $CO_2$ from the early 1960s? Some have argued that the epoch was symbolically launched with the atomic bomb's release over Hiroshima at 8:15 a.m. on 6 August 1945. Naturally, we had influenced both ecosystems and the carbon cycle long before that, but from the post-war period on we became the central driving force behind the globe's development. From an eternal perspective, it matters little whether the Anthropocene started 10,000 years ago, in 1945, or in 1960. We are planted firmly in the human era and our global influence on the biogeochemical cycles is the best proof we have of that. We have quadrupled phosphorus mobilization and almost doubled nitrogen circulation, but our time's greatest headache is the much more modest change to the carbon cycle. The accompanying figure roughly summarizes the earth's annual carbon balance, a short version of a short version.

As a rule, finances are dreary matters for all the non-accountants among us, but everyone recognizes their usefulness. Yet some people continue to put invoices beneath the bed in the hopes they will be spirited away, not unlike our relationship to portions of carbon accounting. As has been pointed out, economy and ecology share many common denominators, despite the fact they have long operated in different worlds. Carbon and climate matters have forced a meeting of disciplines via our relationship to the carbon budget. We must, therefore, tolerate some numbers to get an impression of what is driving what. If we add up the amount of carbon found in ocean, soil, atmosphere and vegetation, we end up with around 42,000 gigatons (Gt) of

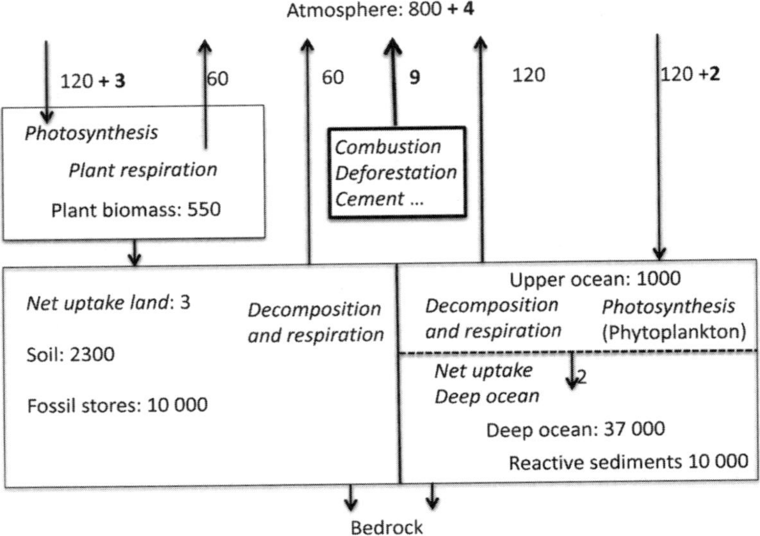


The major aspects of the global carbon cycle. Numbers inside the boxes represents the amount of carbon, and the same holds for the atmosphere, which reflects the amount found in the atmosphere. The arrows denote the estimated annual fluxes, and bold arrows, boxes and numbers represent human contributions. All units are gigatons of carbon.

carbon. A gigaton is $10^9$ tons or a thousand million tons, that is, 1 billion tons. 42,000 billion tons is a lot, but still no more than 0.06 per cent of the carbon permanently locked away in mineral form in bedrock. The carbon initially relevant to us is that circulating in the biosphere in the short (or rapid) cycle, and how it relates to the impacts we make through atmospheric emissions and ecosystem interferences.

Most of the 42,000 Gt C is out of reach in deep ocean areas, totalling about 37,000 Gt. Of what remains within the more dynamic cycle, the division is as follows: land contains a total of 2,200 Gt C, with 600 Gt in vegetation and 1,600 Gt in soil and other dead organic carbon. Every year the land releases 120 Gt C, but it absorbs a little more than that, around 123 Gt, which is about equally divided between vegetation and soil (about 60 Gt C).

The ocean contains about 1,000 Gt C in the upper water layers (where production happens), largely in the form of inorganic carbon: $CO_2$, carbonate and bicarbonate. This stands in exchange with the atmosphere and, as with land ecosystems, the ocean helps us by employing a net absorption that limits carbon's increase in the atmosphere: 90 Gt C is released and 92 Gt absorbed every year. Algae's annual absorption stands at 50 Gt C, that is, the $CO_2$ that enters the ecosystem. Of this amount, about 40 Gt C returns to the atmosphere as $CO_2$ after taking a detour through the food chain, whereas the remaining 10 Gt ends up in the enormous deep ocean carbon reservoir. Most of the carbon here will be respirated by bacteria and vanish as $CO_2$ into the upper water layers yet again.

At present the atmosphere contains 800 Gt C (that amount must be steadily adjusted upwards). This is just a little over what is bound in land vegetation and is significantly less than what we find in soil and ocean. Yet the total output of ecosystems comprises nearly 20 per cent of the atmosphere's carbon content; the most important lesson to draw from this fact is how sensitive the atmosphere's carbon content actually is to changes in the ecosystems' carbon balance. Of our annual carbon emissions of about 9 Gt, land and ocean absorb about 5.5 Gt, so annual growth is 4.5 Gt C.

These numbers, however, are not exact: it is simply not possible to achieve detailed measurements of all carbon fluxes in and out of the world's total ecosystems. We are operating within the correct magnitudes, however, and it turns out that even though an atmospheric 800 billion tons sounds like an overwhelming quantity, it is not actually that much compared to what is sequestered in forests, soil and ocean, and it is certainly not very much compared to what is added and subtracted annually through biological and physico-chemical processes.

Picture the atmosphere as a bathtub. Most bathtubs, at least the slightly older models, have a security drain to prevent overflow if, for example, you happen to be absorbed in a TV thriller while filling the bath. As long as no more water is coming in than the drain can handle, the water level will remain stable, and all you get is a marginal bump in your energy bill due to wasting hot water. Bathtubs hold a lot of water, but if the tap adds more than the drain can handle, you know what will happen – and then you must start talking about costs. Currently ecosystems help by providing a net uptake of $CO_2$. If $CO_2$ levels in the atmosphere continue to rise, it is due to our emissions. This is like equipping the bathtub with an extra tap, not as large as the main tap, to be sure, but enough to exceed the drain's capacity.

There are some caches that do not properly belong to the mineral, plant or animal realms, although coal, oil and gas do originate in the biosphere. Raw oil is also called 'mineral oil', but it definitely has a biological origin that is either animal or plant. Fossil carbon, though, represents only a small amount in comparison to what is trapped, primarily as $CaCO_3$, in rocks. Compared to atmospheric carbon, however, the quantity of fossil carbon is not trivial. The International Energy Agency estimates that the world's extant coal reserves total around 900 billion tons, which, given today's energy consumption of around seven billion tons a year, will last for 128 years.[28] This is considered an 'optimistic' estimate. The World Coal Association's estimate, on the other hand, is only a hundred years of available consumption.[29] Strictly speaking, the word 'optimistic' contains a certain pessimism, since we cannot allow ourselves to consume anywhere near what these reserves contain; at least two-thirds ought to remain securely in the ground – and in thirty years we ought to have abandoned coal consumption altogether.

Currently, China alone is responsible for more than half of the world's coal use, and unsubstantiated rumour has long held that China opens one or two new coal plants every week. Some of these, of course, are intended to replace older, ineffective coal plants, and China performs more responsibly than its reputation in many areas. China is in fact a forerunner in solar energy. Coal is the most carbon-intensive of fossil energy carriers and coal consumption is one of the largest sources of greenhouse gas emissions. Coal, however, is quite diverse and its quality and carbon content depends on where it originated and how it was formed. Brown coal or lignite contains the least carbon (65–80 per cent) and is, therefore, the least energy effective. So-called bitumen (bituminous and sub-bituminous black coal) contains 80–90 per cent carbon, whereas high-quality coal, known as anthracite, approaches pure carbon (90–98 per cent) and is the most energy effective. Roughly calculated, the combustion of a kilogram of coal will produce around 3 kg of $CO_2$.

Regarding oil, BP estimated in 2013 that the world's supply equals 1.7 billion barrels in total, which according to BP analysis is enough for 53 years consumption.[30] These estimates are however in continuous change as new discoveries and new technologies emerge. A barrel contains 159 litres. The combustion of a barrel of oil emits around 320 kg of $CO_2$, but because oil itself is rarely combusted, but rather refined products such as gasoline and diesel, we can say it emits between about 2.3 and 2.7 kg $CO_2$ per litre. Since gasoline's specific gravity is around 0.75 (a little higher for diesel), emissions end up being 1.7 and 2.2 kg $CO_2$ per kilo for gasoline and diesel, respectively, which makes for an acceptable margin over coal's 3 kg $CO_2$. Oil also comes in many varieties, however, and its origin is crucial. Oil from the Amazon is clearly worse than oil from the North Sea, since the former implies the destruction of rainforest and biological diversity, not to mention often heralding the displacement of indigenous peoples, water pollution and more. Canadian oil or tar extraction experiences many of the same problems, whereas oil extraction in the Niger Delta is a terrible industry resulting in a massive oil spill. Oil extraction in countries like Equatorial Guinea is atrocious on purely moral grounds, since the income in this impoverished country lands in the pockets of the kleptocratic government and its closest associates.

Obviously, gas extraction is cleaner all around, but it is even more difficult to give a reliable estimate of the world's total gas reserves. It is also not a simple matter to compare the emissions from a kilogram of gas to those from a kilogram of oil or coal. In contrast, if we look at $CO_2$ emissions per produced energy unit, we come up with a clear range. Coal plainly sweeps into the unattractive first place by a good margin, with higher emissions than oil and almost double the $CO_2$ emission per produced energy unit of natural gas, although this somewhat depends on coal quality. Gas suffers from the problem that it is difficult to transport and a significant quantity can leak out during extraction and shipping. One study of gas leakage covering various fossil industries estimates that global leakage corresponds more or less to Norway's total gas production.[31] Gas may be preferable when it comes to the various fossil energy sources, but the concept of natural gas should not lead anyone to assume it is environmental or 'green' gas. A sign on a gasoline pump labelled 'ecofuel', because of its tiny content of bioethanol made from wood, does not especially make it green. Just because something is less harmful does not mean it is good.

Despite the fact we have ventured some comparisons, it is worth remarking that these same emissions calculations may be presented with completely different answers. This happens especially in estimating current fossil resources and reserves, where the starting point and premises may vary, and it cannot be overlooked that some people will have an interest in talking the estimates up or down. Furthermore, it is notoriously difficult to calculate such quantities. New sources are constantly being discovered owing to improved extraction technology or 'new' types of oleaginous and gaseous energy sources being opened to exploitation, such as oil sand (tar sand) and shale gas.

Added to these considerations are misunderstandings that can yield a range of various answers, such as the fact there are three ways to indicate emissions: as C, as $CO_2$ and as $CO_2$

equivalents. In many ways, C is the most logical since it is C that combusts and C that is released, but still it is $CO_2$ that generates the greenhouse effect. Since C binds two Os, each of which have an atomic weight of 16, the weight ratio of $CO_2$ becomes 12 : (12+16+16), that is 12 : 44. That means that $CO_2$ is 3,667 times heavier than C, and so 1 Gt of C equals 3,667 Gt of $CO_2$. Obviously, there will be a significant difference between using $CO_2$ or C to indicate emissions, since C amounts to only 27 per cent of the weight of $CO_2$. In contrast, when it comes to $CH_4$ there is little difference between $CH_4$ and C, since here C equals 75 per cent of the weight. C equivalents are adopted when gases such as $CH_4$ and $N_2O$ need to be entered into the total greenhouse gas balance sheet. Here it is appropriate to 'convert' them into $CO_2$, because they have 25 times and three hundred times higher warming potential per molecule respectively than $CO_2$. When we want to consider the climatic footprint of some product or service – or of ourselves (as we shall in a final reckoning at the end of the book) – all of our calculations will have a $CO_2$ component, but many will also require us to factor in $CH_4$ and $N_2O$. When it comes to all that agriculture contributes to warming, $CO_2$ equivalents are required since the relative contributions from $N_2O$ and $CH_4$ often far exceed $CO_2$.

If we take the known reserves of coal, oil and gas and convert emissions for their total combustion into $CO_2$ equivalents ($CO_2e$), we get the following calculation: coal, 1,470 billion tons $CO_2e$; oil, 530 billion tons $CO_2e$; gas, 520 billion tons $CO_2e$. The total sum is 2,520 $CO_2e$. Much of our discussion thus far has focused on how long these remaining reserves *can* last us, but the burning question is how long they *ought* to last us. Put another way: how much of our remaining fossil reserves can we risk converting to $CO_2$? There is an answer, but we shall return to it later.

Cycles imply circulation. Much like money, some parts exist in a state of rapid and perpetual turnover, while others might be salted away in the bank or in real estate and remain out of circulation for years. Economists, too, are concerned with economic drivers, interest rates, stock prices and housing prices as they rise or fall. Much is due to psychology, and psychology can transform minor trends to economic catastrophes, for example when shareholders believe the stocks should be sold before the bottom is actually reached. These are self-reinforcing forms of feedback: we all know they exist, but no one, not even the most overpaid stockbroker, knows just where and how they will strike. And then we have adjustment mechanisms, such as thermostatic effects, that change the trend's course. When there is a sense that the trend has bottomed out, brave souls will start buying stocks again, prices will rise, small depositors will follow suit, and with that the cycle is under way again.

Economy and ecology have a surprising number of similarities since, for example, rational actors, optimization theory, game theory and fluctuation analysis are common to both disciplines. Yet there are also some differences. Large economies can actively intervene and mitigate economic fluctuations. While even economic catastrophes can be weathered, the same cannot be said for large-scale ecological catastrophes. We are utterly dependent on nature but not on money. The question, therefore, becomes: how much do we understand about the planet's carbon thermostat and how much can we bank on it?

## GAIA AND FEEDBACKS

It is always a pleasure to trace science's long lineage, to watch how knowledge and inspiration can take us further and bring us to new insights. One discovery that has inspired my own research within so-called ecological stoichiometry – that is, the relative proportion of elements in life and ecosystems – is Alfred Redfield's $C_{106}N_{16}P_1$ (see Part 1). Yet Redfield had long since inspired far greater insights than those surrounding the carbon cycle of a Norwegian forest tarn, spurring James Lovelock towards his theory of the earth as a kind of thermostatic regulatory system where positive and negative feedbacks maintain the planet within certain climatic parameters.[32]

Lovelock is known by some as the man behind the electron capture detector, an ultra-sensitive analysis instrument used to measure the occurrence of different atoms and molecules in gases. This provided indirect support for Paul Crutzen's discovery of chlorofluorocarbon's long lifespan in the atmosphere and its role in the breakdown of the stratospheric ozone layer. Lovelock, in his mid-nineties at the time of writing, is one of those researchers who seem to have a talent for producing new insights and discoveries wherever they turn. One mild May evening ten years ago, as we sat together on an Oslo hotel terrace overlooking the woods and the fjord below, discussing carbon's inscrutable ways, Lovelock admitted that he had constructed the first prototype of the microwave oven, almost in passing, as a practical scheme to use electromagnetic radiation to warm up a quick lunch. The first attempts botched the dose, so the lunch turned to ashes and $CO_2$, but eventually he achieved an output that worked. Others, meanwhile, became officially recognized for the invention of the microwave. Instead Lovelock has been acknowledged for other discoveries, most notably as the father of the Gaia theory.

The background to Lovelock's Gaia hypothesis is one of science history's many examples illustrating the importance of coincidence, this time a letter from NASA that landed in his mailbox in 1961, asking if he was interested in taking part in a research group to explore the lunar surface. This led to Lovelock becoming interested in planetary gas composition, the fact that lifeless planets like Mars have gas compositions in chemical equilibrium, whereas we on earth have a completely different atmosphere. Earth's atmosphere is in thermodynamic imbalance and its gas composition is created and maintained by life. The earth, so far as we know, is unique among planets, not just because it is home to life, but because it has the right conditions for advanced life.

In many ways ecosystems have a thermostatic impact on the earth's climate. Even as they are substantially affected by it, they in turn affect it. This was one of Lovelock's points when he was formulating his Gaia hypothesis, which describes the earth as a living organism. Originally the idea was taken to mean that we could permit our high emissions and impacts without worrying too much about serious consequences. After all, the earth's regulating biological systems would tackle the problem and maintain the globe at a comfortable equilibrium. Lovelock, however, changed his viewpoint because he realized that there were limits to the earth's self-regulative capacity, and that outside these limits we risked everything running amok. His own pessimistic predictions put that level at around 550 ppm $CO_2$. Anything over that, Lovelock argued, risked global warming with a potential to escalate, via a series of self-reinforcing feedbacks, into effects that would shift our planet into unknown territory and result in an inferno the likes of which it has not experienced in 55 million years.

No one, including Lovelock, can be entirely certain if these 'magic limits' exist, and if so, where exactly they fall. In more recent years Lovelock has again begun to doubt his own calculations. Despite that, a 2°C increase in average global temperatures is where the red line is drawn. This 'two-degree target', one of the international community's most important beacons, is certainly not definitive and perhaps even too lax, but it is substantially better than no limit at all. Lovelock's central point, the reasoning behind the two-degree target (actually a limit), is that when life itself is affected, a snowball effect of dramatic and unforeseen self-reinforcing feedbacks could occur within the carbon cycle and the climate. Gaia theory's essential point, meanwhile, is not catastrophic feedbacks taking off, but rather that climate-regulating feedbacks exist.

Gaia theory has taken on various forms of significance. Originally it was connected to the idea that the earth can be regarded as a metaphorical super-organism with the ability to regulate its own temperature, not as effectively as a typical warm-blooded organism, of course, but within the limits that support life. Out of this idea grew a much less scientifically based 'mother earth' conception, which ascribes to the planet almost human traits of care-giving, along the lines of 'Mother Earth knows best'.

Mother Earth, of course, knows nothing at all, but the fact that the globe, climatically speaking, has 'rebounded' following the most extreme catastrophic episodes can give that impression. Similarly, the fact that life has survived every test for almost 4 billion years – even if not exactly the *same* form of life – and the idea that the earth has exhibited an astounding climatic stability through many hundreds of thousands of years has brought relief to all its inhabitants.

It is the case that ecosystems, not to mention life in general, are full of regulating feedbacks on every level. In population biology this idea occurs in simplified form as the fluctuations of prey and predators. The basic point is that an increase in the number of prey animals means more food for the predators who live off them, causing the numbers of predators to increase until predation exceeds the growth capacity of the prey animals, who subsequently begin to decline. This decline will be followed by starvation and decline among predators until the growth rate of prey again exceeds the predation rate. In an idealized world, or in a mathematic-graphical presentation, this concept becomes an endlessly repeating cycle, where predator fluctuations always slightly lag behind those of the prey.

Meanwhile, more than biology plays a role in the various forms of feedbacks. The first compound in Urey's equation, $CaSiO_3$ (wollastonite), and other types of calcium- and silicon-bearing rocks, contribute, as we have seen, to a slow sink of atmospheric $CO_2$. When the earth warms owing to increased $CO_2$, for instance, it will also speed up the hydrological cycle. This will increase the erosion rate, drawing more $CO_2$ out of the atmosphere and then binding it in Urey's reaction, causing the planet to slowly cool again, which in turn slows the erosion rate. A warmer climate and more precipitation, furthermore, stimulates plant growth on the land, which can increase erosion speed by increasing activity in the root zone, partially trapping more carbon in wood and soil. Physical, chemical, geological and biological processes are all at work here, even though a physicist will naturally be inclined to say that everything is essentially physical.

In this type of slow regulation, land upheaval and the movement of continental plates also play a role by forming mountain chains and other steep terrain where erosion and wash-out occur at a quicker rate. Photosynthesis also makes a significant difference. When plants conquered the land during the Silurian period (443–416 million years ago), plant roots helped accelerate the erosion rate.[33] This probably helped cool down the planet; plants, in any case, represented a new adjustment button on the thermostat.

What brought $CO_2$ back – in any case, before the earth's two-legged creatures set in motion the massive combustion of coal, oil, gas and wood – has probably been discharges from deep-sea vents and volcanoes where the Urey reaction occurs in reverse. This slow, cautious erosion thermostat adopts a kind of long-term perspective on things – metaphorically speaking, of course. (No teleology should be interpreted in these regulations.) On the other hand, if we momentarily surrender to the Mother Earth metaphor, we might imagine her spending these days in resignation. If these human creatures keep this party up, the clean-up afterwards will probably require a few thousand years. Still, she has put everything right after ice ages and global hot flashes, so she can probably tidy up the place after humans as well.

Let us hope so, at any rate, though there is not much that can be done in the short term. There are many kinds of chain reaction in the carbon cycle, some of which are self-reinforcing, others of which are regulative. And when it comes to a warming phase, the problematic reactions are precisely those self-reinforcing loops, many of them operative on time scales more relevant to us than the weathering thermostat.

Some will accuse the climate debate of taking place in black and white. Even if that is true, there is an important element of climate regulation that *ought* to have this perspective. Anyone who has ever put their hand on a black and a white surface in the sunshine will recognize the albedo effect. The white surface is cool to the touch, the black surface is warm. Albedo is a beautiful word. It means whiteness (from *albus*, 'white'), and is associated with purity, the sun on new snow and other good things. Nonetheless, I would be less concerned about climate changes if it were not for albedo, or the albedo effect, or more precisely, the effect of *low* albedo. Albedo indicates the amount of incoming light that gets reflected. The antithesis here is the perfectly black surface, which is black because no light is being reflected back. Radiation, namely, is both light and heat. Newly fallen snow can have an albedo of 0.85 (85 per cent of all incoming light is being reflected), old snow an albedo of under 0.5. Ice, as a general rule, has an albedo somewhere between 0.3 and 0.4, which is the same as dry soil or treeless landscapes like tundra and savannah. Forests usually have an albedo of 0.1 (lower for coniferous forests), whereas water surface drops to 0.05, which means that 95 per cent of incoming light and heat is absorbed. Water can appear dark, right? Therefore, the albedo phenomenon is a matter of physics: the question is whether there is snow or not. Yet, the biosphere plays a role here as well, since the landscape with and without forest has a very different albedo. So, why the *fear* of albedo?

It is not actually albedo itself we have reason to fear, but rather its absence. Or, more precisely, a warmer earth means less snow and snow-covered ice in the north, an idea most visible in the dramatic reduction in thickness and prevalence of summer Arctic sea ice. During the winter, of course, the ice is still there, but albedo is irrelevant in the absence of sunlight. Fewer white surfaces mean less sunlight reflected and more heat accumulation in the ocean. Shorter winters also mean that the ground will absorb more warmth. More warmth in the ocean and on land, which could possibly thaw the permafrost, results in less ice and snow, and with that we have a classically self-reinforcing feedback: a positive feedback, technically regarded, but with some negative consequences.

Indeed, northern regions will not just grow warmer and less white, but also greener.[34] Much greener, actually. Just as many Norwegians find their grandparents' cottage, which was built above the tree line, increasingly being enveloped by a forest marching at full speed towards the mountains, so old photos also document treeless tundra increasingly being overtaken by bushes and trees. The phenomenon has also been more thoroughly documented: over the last thirty to forty years, ever more sophisticated satellite observation technology has documented landscape changes on very fine scales. One of the traits measured is the amount of greenness present, termed the NDVI (Normalized Difference Vegetation Index), which captures infrared reflection from chloroplasts. These satellite sensors register reflection from plants – vegetation's albedo effect, if you will. The changes have been strikingly rapid and obvious.

Over the last thirty years, the growing season has increased by an average of six days in the North American Arctic and a full thirteen days in the Eurasian Arctic. In these reasons, temperature is the foremost limiter of plant growth. Higher temperatures and longer growing seasons mean more plant growth. Yet that is only part of the explanation: more $CO_2$ causes a direct fertilization effect, plants simply grow faster, which is further linked to the fact that we have unwittingly fertilized large portions of the northern hemisphere with increased nitrogen emissions from combustion processes and agriculture.

Yet one may well ask, what is wrong with a greener Arctic? Surely more plant life will provide the basis for more animal life and, therefore, richer ecosystems? Undoubtedly, 'greening' has its positive sides, particularly when it comes to increased carbon sequestration. But it also reduces albedo. Less reflection means a warmer surface – and, once again, more vegetation. Historically, of course, there have been significant shifts between tundra and boreal forests. The fact remains, however, that when the forest creeps north due to higher temperatures, albedo decreases, which in turn warms the earth's surface, providing the basis for even more forest growth. The question is, what can prevent this kind of positive feedback from entering a new, irreversible state?

Eventually this process could also lead to a thawing of the permafrost and more $CO_2$ and methane release. Decreased albedo, in short, is only one of many worrisome feedbacks, and in this context the carbon cycle plays a decisive role. Albedo further complicates the calculation of the forest's effect on carbon, simply because more forest means a darker surface with less reflected light. Planting new forests to suck up more atmospheric carbon must be accounted for with lost albedo as a minus.

Carbon has another role to play in this context. As Henrik Ibsen wrote in *Brand* in 1866, comparing the spread of pollution from across the North Sea with the fate of Pompeii:

> Yet worse times; worse visions, frightening,
> pierce the future's night with lightning!
> Britain's coal-clouds spread their gloom
> on our land, foul, black and legion,
> smudging fresh green vegetation,
> spreading vile contamination
> on the fair shoots where it splashes –

stealing daylight from our region,
drizzling down as did the ashes,
once that ancient city's doom. (v:1474–83)[35]

Soot particles created from coal-firing are no new phenomenon and result in darker snow and ice – not to mention a quicker thaw, as anyone knows who has sprinkled ash on spring snow. Soot contributes to there being less ice and snow – and to decreased albedo. And, as we know, soot, black carbon, is a form of pure carbon created by incomplete combustion. Whereas Ibsen, with good reason, directed a complaint against Britain's black coal-clouds, today coal is responsible for only 6 per cent of the world's soot emissions. Today's main source of soot is the burning of forests, peat and grass, whereas biofuel (which is not entirely green itself here) and diesel each contribute around 20 per cent. Soot's health consequences require their own book, but in a climate context soot is not merely albedo's enemy (many blame soot for there being less ice and snow), but itself also helps increase the warming effect. How significant the contribution is can be debated, but some recent estimates place this kind of pure carbon solidly in second place behind $CO_2$ in pure warming effect.[36] This is both good and bad news because soot lasts only a few days in the atmosphere, so cutting emissions would have an immediate effect. In other words, soot is undoubtedly the most low-hanging among the bitter greenhouse fruits.

All this once again shows that 'nature itself' plays a decisive role in the earth's climate, and also how human activity can influence these natural processes and set in motion feedbacks that are practically impossible to control. That is one of the reasons that climate models will always remain notoriously uncertain when it comes to predicting *how* much warmer, wetter and wilder things will get.

And there are many feedbacks out there.

## BLUER AND MORE ACIDIC

One of the advantages to being a researcher, particularly in biology and ecology, is that you get to visit exotic places in an official capacity. I must admit that at one time this motivated my choice of studies, and it has been a source of some memorable experiences – though a forest tarn near Oslo can also be great. Since I have become more conscious of my own personal $CO_2$ budget, there have been fewer of these trips, but you cannot entirely opt out either. Plus, a nearby forest pond does not answer all the questions, such as those concerning $CO_2$'s acidification effects.

On a summer's night in 2010 I was stationed on 'ice watch' and gazed out over a motionless Kongsfjord, in the high Arctic. It was a midnight sun without an iceberg in sight, a rarity at this time of the year, though recent years have been a little different up on Svalbard's west coast. Winter ice has not appeared on the fjord, which might perhaps indicate a warmer ocean. In any case it points to something unusual, but at that exact moment in July, with the large acidification experiment under way, the fact that no icebergs were drifting from the Kongsbreen glacier out onto the fjord was a huge advantage. There is little doubt who would lose in that collision: the nine large plastic containers with their many thousands of litres of water were impressively large, but would offer little competition to a massive clump of ice.

These containers or 'bags' all held 20,000 litres of seawater acidified with different quantities of $CO_2$ in order to explore the effect this had on water chemistry and biology. An entire ship was required to transport the set-up from Germany, not to mention the countless containers with analytical equipment installed in the lab down below. This made it an experiment on an entirely different scale and with very different logistics to the one conducted beneath a tent in the wilderness near Oslo. Each bag was equipped with a different $CO_2$ concentration and the idea was to measure the effect on the ecosystems within the

Some of the large enclosures used during the acidification experiments in Kongsfjorden, Svalbard. Each enclosure had different concentrations of $CO_2$ added and then the effects of water chemistry and biology were monitored.

containers, from bacteria and algae up to crustaceans and sea butterflies (winged snails). The experiment took a month and by the time the whole had been packed up again, we already had many answers since we had done the analysis consecutively. $CO_2$ had certainly done the expected job on the water's chemistry. As predicted, pH sank in keeping with increased $CO_2$ concentrations. This effect is only logical, of course, since we are talking here about basic water chemistry. As usual, it was the biological responses that were more difficult to interpret. Some species have trouble, others remain unaffected. Additionally, we could only measure the short-term effects here. What happens over a few weeks is one thing, over a few decades another.

The acidification mechanism itself is described through the bicarbonate equilibrium, which has four main actors in the carbon family: dissolved, free carbon dioxide; carbonic acid; bicarbonate (the negative charged ion $HCO_3^-$); and carbonate

($CO_3^{2-}$). The relative proportions of these components depend on factors such as water temperature and alkalinity (the water's ability to neutralize a strong acid with a particular pH value). Carbonic acid, or $H_2CO_3$, can protolyse – that is, release protons, or $H^+$, in two steps. The first step frees the first $H^+$ and $HCO_3^-$, the next step frees the other $H^+$ and leaves $CO_3^{2-}$. It is the $H^+$ here that produces the acidifications, and more $CO_2$ will cause more of the bicarbonate reaction that releases $H^+$.

By nature seawater is slightly alkaline with a pH of 8.1–8.2. We are now headed to 8.0 and by the end of this century will probably see an average pH of 7.8.[37] That difference may not sound like the end of the world, but remember that pH is a logarithmic scale. From the Industrial Revolution until the present the ocean's $H^+$ concentration has increased by almost 20 per cent, and while the impact on pH is not yet dramatic (though measurable), there is reason to fear that it may become so. We have to look pretty far back in time to find a parallel to today's acidification rate: some people argue 1 million years, others 55 million years, and still others 300 million. Be that as it may, what we are now witnessing is certainly no everyday occurrence.

Carbonic acid equilibrium is one side of the problem; the other is calcium carbonate equilibrium. $CaCO_3$ is the main component in calcite structures, both among algae and animals. With sufficient $Ca^{2+}$ and $CO_3^{2-}$, it is typical to form $CaCO_3$, though this is also pressure-dependent. With less pressure (in the upper water layers), calcium and carbonate have an easier time staying together. The other variable is pH (that is, the concentration of $H^+$ ions). The ratio of $CO_3^{2-}$, $HCO_3^-$ and $CO_2$ changes with the pH. At a high pH (alkaline), $CO_3^{2-}$ is dominant, at a lower pH, $HCO_3^-$ is dominant, but from a pH of less than 5.5, and even more acidic, $CO_2$ completely overtakes everything.

There is no risk, of course, that the ocean will become *that* acidic, but the problem here is that at a pH of around 8, $HCO_3^-$ is already in quick decline. To complicate the situation further, calcium carbonate has two forms, aragonite and calcite.[38] Different organisms have bet their money on different horses here: some have formed aragonite shells, others calcite. Aragonite has a different crystalline structure and is more soluble than calcite, so organisms that have placed their evolutionary bets here can confirm, in the clear light of hindsight, that this was not a smart choice. The beautiful and graceful sea butterflies belong to this category and they are high on the list of organisms expected to suffer substantial problems. Since they are also key organisms in many northern food chains, this will also be bad news for the many birds, fish and whales that live off them. Corals have also bet on aragonite and what is bad news for corals is bad news for the large number of species that live on or off reefs. Still, it must be said that the Anthropocene, with its human-induced acidification, itself occurring at a historically unprecedented rate, was not easy to predict, and besides, evolution does not plan for the future. Coral reefs, by the way, are not only found in tropical waters. There are also large coral reefs, for example, along the Norwegian coast.

Paradoxically it was in the Arizona desert, so quite some way from oceans and coral reefs, that a start was made on understanding some of the basic connections between increased $CO_2$ and harm to corals. This happened in a place we have already visited, the 'fullerene dome', Biosphere 2. One of the ecosystems it held was a coral reef in an enormous pool. I remember thinking it was just as absurd, and just as impressive, as the rainforest beneath that same dome. As we know, things did not go well with Biosphere 2 after the $CO_2$ level ran amok. When the experiment was abandoned in 1995, most of the fish in the pool were dead and the coral themselves were nearing the end. The marine biologist Chris Langdon was asked to see if this ruined ecosystem could be put to good use in a teaching context. As expected, Langdon discovered that the pH was low on account of the sky-high $CO_2$ level. The negative effect of pH on calcium shells, however, is a function both of pH and calcium carbonate saturation in the water. It had long been assumed that as long as the saturation level equalled one, which meant no net dissolution of calcium shells, everything would largely be fine with the corals. Langdon spent years beneath the Arizona desert dome conducting a series of experiments in which he demonstrated that corals steadily increased their growth to a saturation point (for aragonite) of five.[39]

Until the Industrial Revolution all coral reefs experienced a saturation of between four and five. Increased $CO_2$ levels have now contributed to the fact that there is hardly a place on earth with a saturation level over four, and if developments continue along their crooked path, no coral in the world will experience a saturation of 3.5 (and in 2100 none over three). That does not mean that they will stop growing, but even corals must keep a kind of accounting sheet. Corals suffer a steady loss from grazing, storms and eternal wear, and in order to maintain the status quo, or to increase in size, they must grow rather quickly: the lower the saturation level, the greater the chance that their balance sheet will go into the red.

Lower pH is bad news not only for corals and sea butterflies but for everything containing calcium, from algae to corals, molluscs, echinoderms and fish. The effects on each of these groups can spill over to other ecosystem components, but it is primarily the effects on algae like Ehux that are alarming, since a significant portion of oceanic carbon sequestration is carried out by, in all likelihood, acid-sensitive algae like these coccolithophorids. Less carbon absorption by these algae will result in more $CO_2$ in the atmosphere, which will mean more $CO_2$ in the ocean, which will produce more acidic water, and so on. It is estimated that the current acidification of the world's oceans, a process that shows every sign of continuing, could well produce a situation like the one we experienced 55 million years ago, when temperatures and $CO_2$ levels were also sky-high and the oceans were correspondingly acidic.[40] Thus far, pH and temperatures have remained relatively stable, at least for the last 800,000 years, but we are headed for more stormy waters.

Is it possible to test the acidification effect on a more realistic scale than shock-acidification in bags? Yes, in part you can examine ocean floor sediment from earlier warm periods when the seawater was acidic. The problem here is that only a very limited number of species leave behind significant fossil remains over such long periods, and no one was present back then to measure one-tenth changes in pH. The dramatic warm period 55 million years ago probably experienced an acidification of the same magnitude and rapidity as that upon which we are now embarked. Here the sediments contain a clay layer of dissolved calcium carbonate, so back then life cannot have been too easy for a coccolithophorid or coral either.

There are also natural acidification experiments. Off the Italian island of Ischia are natural $CO_2$-rich underwater vents whose emissions naturally lower the pH by up to 1.5 units – truly, an acid test for what we can expect in a future worst case scenario. Around these vents, otherwise common species of sea urchins, calcareous algae and stony corals are absent and snails along the border zone have noticeably thinner shells. Seaweed and brown algae, on the other hand, benefit substantially from the high $CO_2$ (not to mention the absence of grazing urchins) and so thrive in the acidic water. Truth be told, the pH here is still between 6.5 and 7.0, so not terribly acidic in a chemical sense, but it is still enough to create problems for the formation of calcium shells.[41]

Another feedback concerning algae is the possibility that fewer nutrients will make it to the water's surface. This effect can happen if surface warming contributes to a stronger temperature gradient or, to put it another way, if temperature differences between surface water and deep water grow greater. That will mean less stirring, less nutrient-rich deep water making its way to the upper water layers where algae can utilize it. If the ocean absorbs less $CO_2$, furthermore, more of it remains in the atmosphere, it becomes even warmer and so on. Many studies show a reduced quantity of chlorophyll in the world's oceans and this process might be part of the explanation why. Less algae has not only the unfortunate effect of binding less $CO_2$, but results in lower production on all levels of the food chains. In short, we may see more of the clear, blue water that is characteristic of low-productivity ocean areas, but in this case clearer water does not bode well.

The acidification feedback can be summarized as follows: more $CO_2$ in the atmosphere means more $CO_2$ dissolved in the water. This causes the water to become more acidic, which contributes to less absorption of $CO_2$ by the ocean, which in turns harms algae. As such, $CO_2$ concentrations increase more rapidly, more $CO_2$ is dissolved, the water turns more acidic – and here we go.

Still, we have many feedbacks. To repeat an important point: it is life processes that shape – and maintain – liveable conditions here on earth. Ecosystems both release and sequester $CO_2$, some ecosystems (like most inland lakes) emit more $CO_2$ than they take up, a fact also true of certain ocean areas, but in general we see a net carbon sequestration – certainly in the boreal coniferous forest. However, ecosystems also produce methane ($CH_4$) and nitrous oxide ($N_2O$), something true of natural ecosystems, agricultural areas and agricultural activities in general. Fertilization can result in increased nitrous oxide production and cows are notorious methane sources that contribute significantly to the global methane budget. Cow farts, though, are not the greatest cause for concern.

## METHANE BOMBS

Warmer northern summers mean that snow cover will shrink and the earth's surface will darken. Because of this it will get even hotter, perhaps so warm that the massive, northern carbon reserve will begin to stir. North of the boreal coniferous forest belt is a nearly treeless tundra. A tundra is typically a frozen expanse, only the uppermost soil layer of which thaws in the summer. The tundra contains vast quantities of frozen carbon and methane and, in a climate context, one nightmare vision is that this colossal reservoir will thaw and emit these frozen greenhouse gases. When and if this will happen, and exactly how much $CO_2$ and $CH_4$ we are talking about, remains extremely uncertain. The first question to ask, of course, is whether the permafrost is actually thawing. We have seen photos of crooked trees and buildings in Alaska, caused by disappearing ground layers as the frost vanishes. We hear tales of more frequent discoveries of mammoth teeth (sometimes whole mammoths) as the permafrost thaws and the ground gives way in Siberia, but just how systematic and comprehensive is this process?

Some places are undoubtedly experiencing a thaw, but in saying that it is important to keep in mind that this is nothing like thawing the home freezer. This ice contains an enormous freezing capacity, and potential melting depends not only on temperature changes but on snow cover. Nonetheless, the process will be really slow. There is little doubt a thaw could release vast quantities of greenhouse gases. Simultaneously, increased plant growth in the tundra could help bind a portion of this extra carbon. Perhaps the tundra's fate is indeed to play the wild card when it comes to climate development.

Swamps, wetlands, lakes and oceans release large quantities of $CO_2$, $CH_4$ and $N_2O$. Whether this contribution will increase under a changed climate remains unclear. Especially in terms of methane, this depends on whether the climate becomes drier or wetter. Warmer and wetter conditions will mean a higher 'natural' output of $CH_4$, whereas a drier climate will reduce $CH_4$ output but perhaps give us more $CO_2$. Nitrous oxide emissions, in contrast, are largely connected to the nitrogenous fertilizers we use, to combustion processes and to changes in the nitrogen cycle.

Coal and oil deposits generally contain methane as well; it is not without reason that methane has always been one of the scourges of coal mines. Oil extracted from the depths can sometimes bubble with trapped methane as the pressure decreases – like opening a bottle of fizzy drink. Methane emissions from gas production, however, loom the largest, and in its fifth report the IPCC pegs methane as the greatest unknown in the climate equation. Many people believe that natural gas extraction, and especially shale gas, produces significantly more emissions than we have recognized (or wanted to acknowledge). A new U.S. study indicates that methane emissions from American gas production are underestimated by 50 per cent and gas leaks out from most fossil recovery processes. If a notable gas leak occurs on top of this, suddenly gas is not nearly so climate-friendly (compared to coal and oil) as has previously been claimed. Frozen gas hydrates have also attracted interest as a potential energy source. The problem here is separating methane from ice, something that can be done either by reducing the pressure or injecting $CO_2$ into the hydrates so that water molecules will bind to $CO_2$ and release the methane. The greatest challenge here is carrying out the process without too much 'mess', thereby contributing to what many fear will happen with warmer oceans – namely, that the methane in gas hydrates will end up in the atmosphere.

We recognize the broad outlines in the carbon cycle, and the types of responses, but our knowledge has many unknown sides. For example, we know dangerously little about the durations and consequences of the changes we are now setting in motion. One of the greatest unknowns here is methane's role. As we have seen,

$CH_4$ is connected to $CO_2$ in many ways. Organic carbon formed from $CO_2$ during photosynthesis forms the raw material for most biologically produced methane, and much $CH_4$ is 'consumed' and transformed to $CO_2$ by bacteria. They also enjoy a common fate in other ways. Historically, concentrations of $CO_2$ and $CH_4$ in the atmosphere fluctuate at the same rate and these fluctuations keep pace with the temperature. Warming caused by $CO_2$, whether due to our emissions or to other sources, could unleash a significant increase of $CH_4$ and that also applies to other situations.

When it comes to $CH_4$ emissions, there is little reason to expect any dramatic changes in the world's termite populate (or livestock for that matter), whereas methane hydrates, which today contribute only an estimated 1 per cent of total emissions, pose a climatic wild card. The question is always how much of an alarmist you want to be, but it would be quite irresponsible for a doctor who, upon discovering a potentially fatal tumour, failed to inform the patient. Everyone agrees that gas hydrates contain enormous quantities of methane, but no one knows just how substantial these stores are or how securely they are kept. There is probably much more there than the 15,000 billion tons (15,000 Gt) that so often appears as a 'best guestimate', since caches of non-biological, that is to say, chemically formed gas hydrates can be considerable.[42] How much of this reserve will potentially make its way into the atmosphere is another thing – the least amount possible, it is to be hoped.

It is probable that gas hydrates harbour a warming potential that surpasses all the known oil and gas reserves combined. Five hundred gigatons of carbon in the form of $CH_4$ is equivalent to a tenfold increase in today's $CO_2$ levels in terms of warming effect.[43] This is a nightmare scenario, but luckily an improbable one. Most of these deposits appear to be quite safely tucked away at great depths, unless someone truly discovers a way to shake the globe. The exception is methane hydrates found in the relatively shallow waters off Siberia and northern Norway, as well as what might be located in the permafrost. We do not expect some kind of spontaneous methane release, but we could see a steady increase with warmer oceans (temperatures in Norwegian coastal waters have already risen by almost 1°C) combined with permafrost thaw.

Bacteria, of course, have not always governed the methane cycle. When our planet was still in its infancy 3.5 billion years ago, $CH_4$ concentrations were probably around a thousand times higher. Back then the numerous and active volcanoes were to blame. It was around the same time that the first arch bacteria began stirring in the primordial ocean and they also began producing $CH_4$ on the basis of $CO_2$ and water. Because the atmosphere lacked oxygen, there were not too many hydroxyl radicals around to keep the methane concentration in check.

Why then was the earth not roasted to Venus-like conditions by the greenhouse effect back then? Certainly, the atmosphere was richly supplied with $CO_2$ around that time. Of course, Venus does not have much in the way of methane, but has way more $CO_2$: 96 per cent – which is much of the reason the planet has an average surface temperature of 462°C. Earth was probably a lot hotter back then, but the sun was significantly weaker, and the heat was not so unbearable that it prevented blue-green bacteria and eventually others from producing enough $O_2$ that methane breakdown outpaced its formation, causing the methane concentration to sink to today's familiar levels.

The fact that methane concentrations during the planet's early years, however, were significantly higher than they are today is poor consolation. Judging by the analysis of gas concentrations in ice cores, $CH_4$ is now double what it has been at any time during the last 420,000 years (which is as far back as reliable measurements are possible).

SINCE OUR own $CO_2$ and $CH_4$ emissions into the atmosphere are relatively modest, they have long been argued about by fossil fuel lobbyists. Why fret about our own modest $CO_2$ emissions when nature is responsible for the remaining 94 per cent? It is here the worry lies, however, since not only does the earth's own exhalation balance its corresponding inhalation, but its inhalation exceeds exhalation such that carbon sequestration in ecosystems has also handled a significant portion of our own emissions. When this capacity is reduced, and potentially combined with increased release from the vast carbon stores on land and in the ocean, we are headed into terra incognita.

The critical question becomes whether these feedbacks will lock the planet into an irreversible, one-way process. The fact that gases in the atmosphere have remained reasonably stable for hundreds of thousands of years, despite their much shorter atmospheric lifespan, shows there must be mechanisms in place that help maintain this stability.

Ecosystems themselves are notoriously unstable in the sense that they undergo changes due to seasons, precipitation, warm and cold years, species varying in number and predominance, and

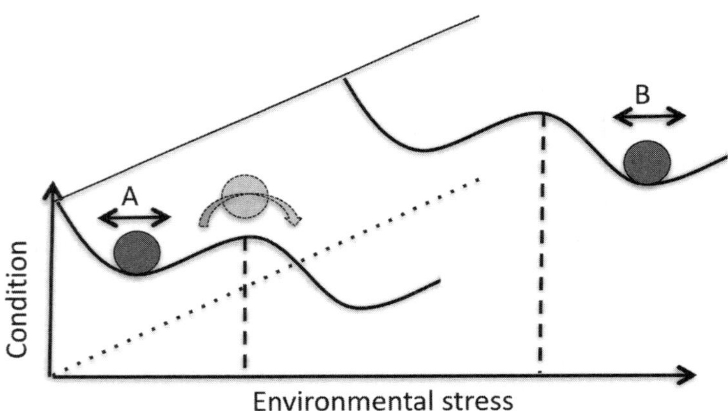


Many ecosystem responses are characterized by thresholds, meaning that up to a certain level of environmental stress the system is locally stable (the ball remains within the cup (A)). Above a certain level of ecosystem stress, however, the ball may be pushed above a threshold, and the system enters another state (B), from which it is hard to return to A. Could this happen on a large scale, affecting areas like the Amazon, or even the global climate?

so on. Despite this fact, they are often what we term dynamically stable, with variations happening within logical boundaries, as we can illustrate with a ball in a cup. If we change conditions enough, the system enters a critical level where the ball rolls into a new state of equilibrium. Such system changes can be both local and global. The earth has probably undergone such changes of state, but has never spun completely out of whack. There have always been mechanisms to roll it back into place within the hollow where the planet is still a paradise, at least climatically speaking and with a slightly broader view.

## BETWEEN A SNOWBALL AND HELL

A common argument throughout this book is that even though our knowledge of the carbon cycle is fraught with gaps and deficiencies, there is a form of overall regulation. This is not conscious regulation, of course, but a network of short and long cycles, respiration and photosynthesis, weathering and precipitations that apparently turn down the thermostat when it gets too hot and turn it up when it gets too cold. Yet, how might such a thermostat tolerate sudden changes, like someone turning the heat on high with the windows closed? No one knows, but we logically hope that the thermostat will be finely tuned enough that not only bacteria and cockroaches will survive, but that conditions will be liveable from a human standpoint as well.

What we do know is that, despite ice ages and warm periods, conditions have remained sufficiently stable for life to persist and evolve. Concentrations of $CO_2$ and $CH_4$ have fluctuated up and down a bit in keeping with temperature, rather like ripples. Farther back in time conditions were more turbulent, but life has continued unbroken through great and small tribulations since its humble beginnings more than 3.5 million years ago. There have still been some close calls. The worst was around

600 million years ago, since much evidence indicates that the planet was a hair's breadth away from a complete freeze during an ice age like no other.[44] There have also been times when life sweated it out in the greenhouse, yet still managed to get along. All this is simply to illustrate the fact that we are not the only species capable of dramatically affecting the earth's climate, in case that point was in any doubt.

The fact that $CO_2$, $CH_4$ and temperature fluctuations remain so remarkably in step has given rise to heated discussions regarding 'the chicken or the egg'. Are we now witnessing a warming that, as a secondary effect, causes ecosystems to emit more $CO_2$ and $CH_4$? Or is it the case, as climate researchers tend to argue, that increased human emissions of these gases cause temperatures to rise? Because gas concentrations and temperature shifts are so closely linked in time, it is difficult to distinguish the chicken from the egg in ice core measurements and sediments. Temporal resolution is just not good enough to provide a clear answer, although newer studies do definitively point towards gases increasing first. And what else could increase temperature? A good answer here is conspicuous by its absence. That does not mean, of course, that earlier periods have not seen the opposite arise. Orbital tilts, solar activity or geological processes could all have increased temperature before gases. This close connection is alarming, since it could be an expression of the feedbacks we have examined. More gases, warmer climate, more gases . . .

Since the last ice age the globe has been remarkably stable. The global average temperature has generally varied by less than 1°C, even when we include the warm period that reached its maximum around 1,000 years ago, as well as the Little Ice Age in the 1600s. Further back, fluctuations have been significantly greater, usually spaced about 100,000 years apart. The longest time series from the Russian station Vostok in the Antarctic reaches 3.6 km into the ice and covers a period of almost half a million years and four distinct ice ages.[45]

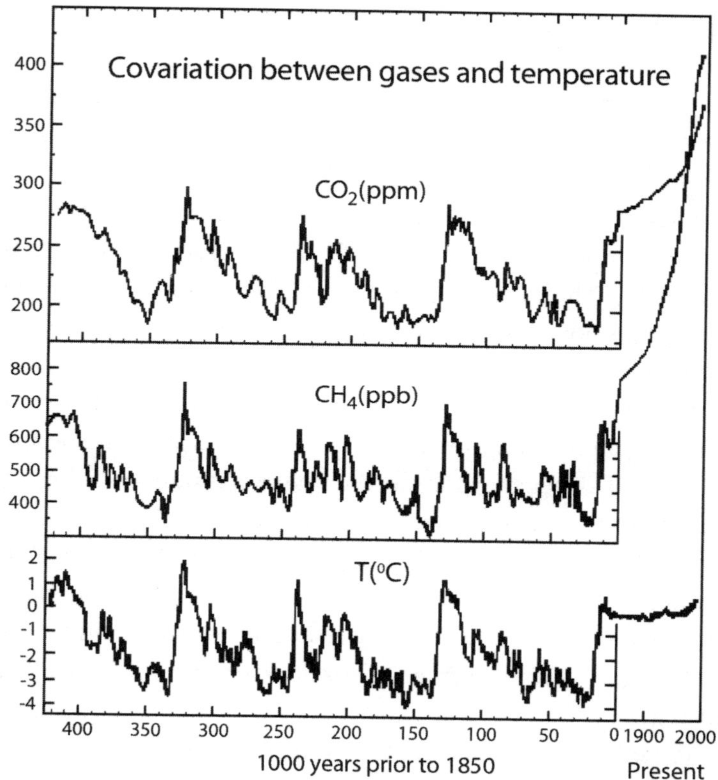


Reconstructed temperature and atmospheric concentrations of $CO_2$ and $CH_4$ from the Antarctic ice sheet. From the Vostok series that extends more than 420,000 years back in time.

One theory regarding periodic climate fluctuations was proposed by the Serbian mathematician Milutin Milanković in 1916.[46] Because the earth's orbit is elliptical, there will be periods of stronger and weaker insolation (incoming solar radiation), and Milanković calculated a cyclicity of 23,000 and 41,000 years, respectively, where the earth's axis, its angle of tilt with respect to the sun, and its speed of rotation are figured in. The effect was that solar intensity on the earth varied at regular intervals. If historic climatic variations are due only to insolation effect variation, suddenly the earth's thermostat is no longer quite so impressive, but is more like the presence of an invisible hand turning the heat lamp's intensity up or down. Milanković's calculations are logical, but correspond poorly with the roughly 100,000-year observed cycles. Can there be other mechanisms helping the earth recover itself again?

Even these 100,000-year cycles seem to be subject to still longer fluctuations, but here the historical archive begins to be less reliable. The ice archive, in any case, can no longer help us (ice clearly does not preserve prehistorical temperatures directly, but rather atmospheric isotope signals, which are just as dependable). There seem nonetheless to have been some episodes when the earth was on the verge of reverting back to the place life began.

PETM (Palaeocene-Eocene Thermal Maximum) occurred 55 million years ago, and if we really want to drive the point home and explain just how bad it can get if we 'stay the course' for decades to come, we usually refer to PETM.[47] Although the causes then and now were different, the temperature effects and ocean acidification were the same as those towards which we are now veering.

PETM is documented by the isotope ratios in $CaCO_3$ contained in deep ocean sediments. These show that $CO_2$ increased dramatically and that temperatures also increased 'rapidly' (in geological terms) by 5–10°C. After that it was some 200,000 years before the system recovered, but PETM marked the dramatic beginning of a troubled epoch that experienced several abrupt warming episodes. These were not on a scale rivalling PETM, of course, but they still led to a 3°C increase in average ocean temperatures, a formidable ocean acidification and a number of species vanishing for good.

For anyone anxious about the planet, it is naturally not comforting to know that such upheavals can occur, but still there is a certain consolation in the fact that even PETM did not prompt an irreversible, one-way process with extreme emissions of $CO_2$, $CH_4$ and temperatures barrelling in the direction of Venus-like

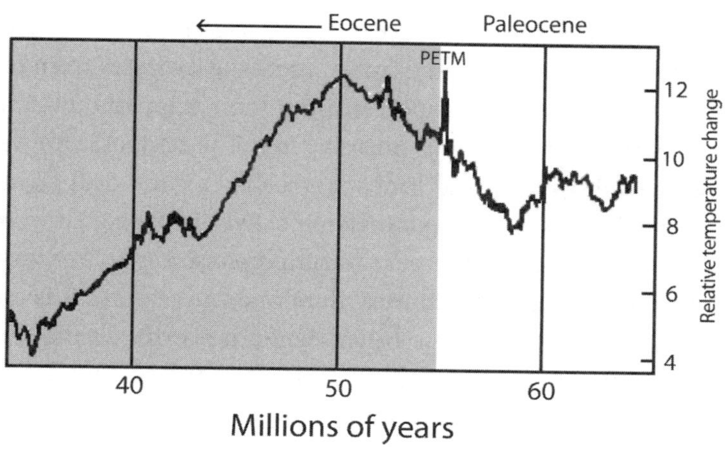


Estimated temperature development around the PETM event.

conditions. One difference between Venus and Earth, of course, is that Venus has lost all its water. Without water there can be no weathering, and weathering is a key factor in $CO_2$'s removal from the atmosphere. On Earth, at least, it seems that the thermostat(s) have slowly but surely brought things back to working order.

What unleashed PETM? No one knows for sure, but $CO_2$ and $CH_4$ must have been involved, though it is not clear what came first. Both doubtlessly compounded the situation through a series of feedback effects. The supercontinent Pangaea began to break up 200 million years ago due to continental drift and the separation was nearly complete at the end of the Palaeocene 55 million years ago. When the land masses in the North Atlantic separated, they literally opened the way for massive volcanic outbreaks and the atmosphere was pumped full of $CO_2$, perhaps enough to raise the temperature by a couple of degrees – the current path we also are on. Here is where the story gets frightening. This temperature increase seems to have stimulated even more $CO_2$ emissions, because of more acidic oceans, for example, and perhaps also a net release of $CO_2$ on land. And the temperature rose. The next step in the process might have been that methane really began to stir, with the frozen methane hydrates from the ocean floor thawing out and releasing their trapped methane, thereby throwing more gas onto the fire. The temperature was now headed over 5°C and probably reached a good deal higher than that before evening out, when it slowly sank over the next 200,000 years before the next warming phase set in. Yet what triggered the methane hydrates' thaw and gas release? It is difficult to imagine any force other than a powerful earthquake and volcanic activity that could rupture the earth so completely. Another explanation might be the melting of deep ocean rocks with a sizeable organic carbon content, which in a short period of time were then 'vented' through the ocean floor.[48] No matter the cause, the truly *terrifying* fact is that even during PETM, we are not talking about annual $CO_2$ emissions that exceeded 5 billion tons: that is only a seventh of today's emissions.

Other things happened during PETM. Under high temperatures in the depths, some of the organic carbon in the sedimentary rocks boiled out as $CH_4$ and $CO_2$, which contributed to further warming, but it is also highly possible that some of that organic material under high pressure became oil and gas (methane). The gradual leakage of methane likely contributed to persistent high temperatures for a long time afterwards, even though the extreme peak was reached during PETM.

Yet not even PETM was the end of the world; life in the broad sense dragged on. This was not necessarily the same kind of life, but most life forms appeared to have made it, perhaps by migrating to the poles where temperatures were still cooler. Likely Urey's reaction brought things under control again, but it took several hundred thousand years. By that time, plants, fungi and bacteria truly had colonized the land, so these also helped strengthen the weathering process, while forest and ocean worked their biological carbon pumps from the atmosphere to soil and sediment – as they continue to do today.

PETM and other warming episodes in the Eocene (the period from 56 to 34 million years ago) have not been the only ones to confront the globe. The earth has always been subjected to larger and smaller catastrophes, though what is catastrophic opens possibilities for others. The increase in oxygen was a slow catastrophe for the old, anaerobic life forms, but also the basis for all 'modern' and more complex life forms.

My colleague Henrik Svensen is among those to have delivered convincing indications regarding the way substantial gas emissions ratcheted up the temperature even before PETM.[49] At the start of the dinosaur era, on the Permian and Triassic boundary, a dramatic warming of near-PETM dimensions occurred. The cause was probably $CH_4$ and $CO_2$ vented from huge, chimney-like volcanic structures in Eastern Siberia. The earth's temperature increased an estimated 6°C on average; the oceans became acidic and, even worse, substantially oxygen-free. This was one of the most dramatic mass extinctions in life's history, and it proved especially devastating because the oceans became more acidic at the same time as deep ocean oxygen content was reduced. There was also an emission of ozone-depleting substances into the atmosphere via those 'pipes' in Siberia. At that point, it also took a magnitude of 200,000 years before temperatures normalized, though the oceans needed half a million years to recover. Forests evidently struggled longer, since there is a period of almost 10 million years in the early Triassic with no coal deposits ('the coal gap').

Perhaps this spawned the age of the dinosaurs; no one knows. What we think we do know is that a massive meteorite put an end to the dinosaurs about ten million years prior to PETM. The dinosaurs would have probably made it through PETM, at least some of them, but the meteorite, perhaps in tandem with volcanoes, reduced plant growth and cooled the planet by blocking out the sun. Reduced plant growth, of course, gradually meant more atmospheric $CO_2$, which again made conditions liveable, at least for the side-branch in the dinosaur family tree that did not snap off in the storm: birds.

This temporary cooling, however, is nothing compared to the aforementioned freeze, which is known as 'Snowball Earth', 650 million years ago. What precipitated this catastrophic freeze is debatable. Some argue that the snowball concept is hyperbolical, but still there is much to indicate that periodically the planet became frozen so completely that ice began to creep towards the equator. If ice from the north and south had met at the equator, it would have been a dark picture indeed amid all that white, although it is entirely possible the thermostat would again have demonstrated its capacity. Volcanoes would have continued to spew gas into the atmosphere, richly supplying it with $CO_2$, and with a frozen surface the weathering drain would have been defunct. As such, $CO_2$ could gradually build up in the atmosphere to levels where it would warm enough to thaw the ice, the earth's surface would emerge, and we would be under way again. It is perhaps this mechanism that kicked in 600 million years ago, and maybe also with previous freezing episodes, so life had an equatorial safe haven until larger portions of the planet were once again liveable.

Weathering's slow $CO_2$ capture would be robust enough to keep pace with slow changes in emission. With the exception of truly colossal eruptions, it is also sufficient to control $CO_2$ emissions from geologically active regions. When catastrophes do occur, the weathering equilibrium does get a handle on things, but not with any haste. In terms of the big picture, 100,000 years is not all that much – the globe, after all, is 4.5 billion years old – but for us this is no consolation. We need something that can twist the thermostat a little more emphatically, and it is here that biology comes to our aid, at least until we ratchet up the heat so much that the thermostat simply takes a pass.

As we have seen, forests help accelerate the weathering rate and forests themselves are climate-dependent. Increased weathering sequesters more $CO_2$, thereby lowering atmospheric $CO_2$ levels. As such, the climate cools and there is less precipitation, less forest growth and reduced weathering, until $CO_2$ production again outpaces the weathering rate. But other things happen during weathering: phosphorus and silicate are released from the bedrock, some of which travels via rivers to the ocean, where it stimulates production. This also increases $CO_2$ uptake through higher algae production. Remember that the ocean's phytoplankton bind almost as much carbon as land vegetation does, though otherwise they play very different roles. Algae contribute nothing to weathering, at least not directly, and therefore play no role in the weathering thermostat. In contrast, algae form a crucial basis for nearly all ocean production through the food chains and also function as the biological pump for transport not only of carbon, but of other important nutrients, such as phosphorus, nitrogen and silicon, from the productive upper water level down to the dark ocean depths. In this process Ehux and other coccolithophorids play a key role, not to mention the siliceous algae that often prevail during springtime algae blooms. These latter algae are what mainly profit from an increased supply of silicate ($SiO_2$), since it forms a necessary building block in their silica shells, but they also contribute to a continual depletion of silicate from the upper layers into the depths.

Algae also contribute in another way. Many groups produce dimethyl sulphide, which we discussed earlier. This compound can have a number of functions and is probably important for algal osmotic regulation. Significant quantities of dimethyl sulphide end up in the atmosphere, thereby becoming an important source of sulphate aerosols, which are again important for cloud formation, thus impacting the climate. This sulphate effect was a central inspiration for Lovelock's Gaia theory. More algae not only remove more $CO_2$ from the atmosphere, but can have a cooling effect as more sunlight is reflected from clouds. The reduced weathering rate, which follows from less atmospheric $CO_2$, will also contribute to less phosphorus and silicate being transported to the ocean.

There are many other such feedbacks, some relatively well documented, others probable, still others more speculative. Lower algae production on account of reduced phosphorus and silicate will contribute to lower oxygen concentrations in the ocean, increasing the risk for oxygen depletion in the depths. The same thing might happen with higher ocean acidification. This will smartly affect the nitrogen cycle, thereby regulating the nitrogen supply for algae production. More nitrogen available might incline the system towards a phosphorus limitation (as explained by Liebig's barrel staves, discussed in Part 1). With phosphorus as the limiting factor, weathering will become critical for ocean production.

We could continue discussing such complex interactions and feedbacks, but we will leave off here. The point is clear: these cycles overlap like cogs in a machine. It is, though, a machine we do not entirely understand, while the cogs vary in size and belong to differently geared mechanisms. In order to build a cell or an individual, carbon must unite with its good neighbours on the periodic table, especially nitrogen and phosphorus, but also oxygen. In the same way, the global cycles are linked through a common destiny. Thus far in our discussion, oxygen has appeared as carbon's sidekick, though this partnership obviously has global consequences. Since oxygen is undoubtedly the element with which carbon tends to form the most intimate relationship, even though they are in many ways opposites (while noting that opposites attract), it is time to take a closer look at oxygen.

## OXYGEN'S EVOLUTION

Oxygen has always been carbon's soulmate, and not only because together they form $CO_2$, which wields such power over life, but because oxygen and carbon mirror each other in so many reactions. Even though the atmosphere holds immensely more $O_2$ than $CO_2$, there is more than enough organic carbon in the ocean and on land for these two elements to influence each other mutually. Atmospheric oxygen also has its feedback mechanisms, many of them closely connected to carbon.[50]

For perhaps as much as 1 billion years, the atmosphere remained virtually oxygen-free after the first blue-green bacteria began the laborious task of filling the atmosphere with it. Admittedly that task was completely accidental. Oxygen was primarily a necessary evil, a by-product of the bacteria's learned trick of splitting water as the basis of photosynthesis. How do we know this is all true, in an epoch that left neither fossils nor frozen air?

In the absence of oxygen but in the presence of iron and sulphide (such as in $H_2S$), iron disulphide ($FeS_2$), a substance also called pyrite, will form. Everyone who has gone into a geology museum's souvenir shop with a child has ended up buying a piece of iron disulphide, which fortunately does not smell like hydrogen sulfide. Sulphur paired with iron has instead created beautiful brass or gold-coloured cubic crystals – something you might call a fortunate partner swap. Since iron sulphide can form only in oxygen's absence, you can trace geological layers containing pyrite-rich alluvial deposits up until 2.4–2.3 billion years ago. At that time, traces of iron disulphide and other easily oxidized compounds suddenly disappear from the sediment, indicating $O_2$ has made an appearance. This evidence is supplemented by analysis of sulphur and carbon isotopes from other same-aged geological occurrences, where comparative indications show that $O_2$ suddenly passed a threshold value. It would be logical to conclude that oxygen's increase happened at a slow and steady pace, but in this case nature seems to have departed from Darwin's rule 'Natura non facit saltum' (nature makes no leaps). Of course, Darwin was talking about evolution, but here we nonetheless see a leap of nature, primarily as a result of sudden changes in the atmosphere's oxygen content.

Oxygen back then was still far below today's levels, but the sudden changes indicate that there must have been a shift in the relationship between produced and consumed $CO_2$. The producers were largely blue-green bacteria, but there were many other sinks. Most of the oxygen went into chemical oxidation processes, for example the oxidation of pyrite to sulphuric acid, but around 2.3 billion years ago the balance sheet documented a plus. At the same time, the quantity of $CO_2$ decreased as a result of photosynthetic carbon sequestration.

Here we come to an interesting feedback. More oxygen signifies the oxidation not only of iron disulphide, but also iron sulphide (FeS), resulting in sulphuric acid rainfall, just like happened during the worst acid precipitation period in the 1970s and '80s. At that time, the sulphur in question originated from the coal-combustion processes then employed, sulphur that today we rinse away. Yet no matter the source, sulphuric acid will promote increased weathering, which in turn releases more phosphorus from the bedrock, especially from the enormous stores of phosphorus bound as apatite, a mineral not unlike our teeth and bones. This feedback is only a hypothesis, but a logical one. If it is correct, increased $O_2$ in the atmosphere resulted in acid precipitation, which in its turn yielded more phosphorus (and probably more silicate), which again led to a formidable increase in ocean production, which increased the atmosphere's $O_2$ content (at the expense of $CO_2$).

Paradoxically, this acidification could have contributed in the long term to improving the ocean's buffer capacity since more $CaCO_3$ was being released from rocks. The same will be true of modern, human-induced acidification relating to sulphur fallout from coal plants. While this has led to lakes and rivers acidifying, it has also supplied buffering $CaCO_3$ to the ocean, thereby helping combat today's $CO_2$-driven acidification. We actually see the calcium content decrease in many regions at the same rate as acidification decreases. The reduction in acidification is obviously great, but not so much the reduced Ca. Many freshwater species risk a type of osteoporosis as they develop problems with their calcium shells, not because of acidification but due to the lack of building material. Meanwhile, the ocean is supplied with less buffering bicarbonate. Here we can again play the pessimist and say that one person's good is another person's ill. This may be speculative, but is it at least plausible.

It still took quite a while, of course, for oxygen to reach current levels. Eventually the atmosphere reached 20–25 per cent $O_2$ and it has probably stayed there for the last 500 million years, maybe longer. More complex, multicellular molluscs appeared 575 million years ago, and 25 million years later saw the 'Cambrian explosion'.[51] It is an explosion only through evolutionary lenses, but it represents a period of a few million years in which life experimented. Numerous new animal forms, a characteristic group being the trilobites, appeared and disappeared, representing yet another counter-argument to Darwin's perception that life's development must take place gradually. Since we are talking here about animals that, like us and all other higher life forms, were dependent on respiration (cellular respiration with oxygen), there is good reason to assume that it was blue-green bacteria and other photosynthetic organisms that created the basis for higher life here on Earth. Furthermore, it was only with the creation of a modern atmosphere that the ozone layer could form (ozone consists of three oxygen atoms, $O_3$), serving to filter out ultraviolet radiation's most shortwaved and harmful component.

Hundreds of millions of years after the Cambrian explosion, $O_2$ began seriously to increase. The immense Permian and

Carboniferous swamp forests ruled the globe, forming the basis for most of the coal, oil and gas that were to be burned in a historical blink 300 million years later. Massive oxygen production was under way, and at its peak the atmosphere contained perhaps 35 per cent $O_2$.[52] It is from this period that we find fossils of monster insects, our first hint that something unusual must have happened.

Between 1877 and 1894 the palaeontologist Charles Brongniart conducted studies near Commentry, central France, in a quarry that had once been a lake bottom and provided geological evidence going back 300 million years.[53] It is often lake sediment with oxygen-free bottom water that gives us the best fossils since dead animals and plants receive a gentle burial in the muck here and do not decompose to any great extent. When the mud at the bottom is compressed and hardened into rock, which then undergoes an upheaval, perfectly preserved fossils can see the light of day at random, perhaps at the stroke of a palaeontologist's hammer.

Brongniart's hammer struck just such a treasure trove. His most famous find was the dragonfly *Meganeura*, which has a wingspan of 63 cm and today can be viewed at the Musée National d'Histoire Naturelle, Paris. This inspired others to join the fossil hunt, which yielded hummingbird-sized caddis flies, metre-long millipedes, spiders that put today's tarantulas to shame and many other giants. As recently as 2005 traces were found of what was believed to have been a 1.5-m-long water scorpion in a geological site from the same period in Scotland (today's water scorpions are about 1.5 cm). On land insects and arthropods have an ineffective method of providing their body with oxygen. Since they lack lungs, oxygen must be diffused by air ducts located in the body (trachea). The larger the insect, the more difficult it will be to supply its body with oxygen in this way. Record high $O_2$ levels must have temporarily overcome that problem.

The heyday of monster insects eventually came to an end and the pendulum swung the other way as $O_2$ plummeted to

15 per cent. The thermostat kicked in, but not in any precisely regulated way. The massive plant production on land with its subsequent storage of coal, oil and gas must have taken a toll on the atmosphere's $CO_2$. On top of that, a huge weathering of land plants, which we have previously discussed, provided the ocean with phosphorus, further reducing $CO_2$. Eventually plants must have been struggling for breath, strangled by their own success, gasping after $CO_2$ in all that oxygen.

Oxygen's many links to and feedbacks with carbon are in part slow and complex, in part indirect and purely speculative. However, there is one instantaneous and direct feedback that concerns both elements and which we ourselves can observe: forest fires. Fire has had, and continues to have, a substantial effect on the earth's carbon budget and plant development. It is estimated that 4 per cent of the world's vegetation-covered areas burn every year.[54]

If the atmosphere's $O_2$ concentration reaches levels significantly higher than 30 per cent, the risk of forest fires will increase. At even higher concentrations, a simple lightning strike could be enough to start a fire. Massive forest fires produce massive $CO_2$ emissions, and even though this would not make a dent in the atmosphere's $O_2$ content, it would still notably contribute to reducing land vegetation's $O_2$ production. Reduced forest cover will again slow the weathering pace, the transport of phosphorus and silicate to the ocean will decelerate, and algal $O_2$ production will also sink. Three hundred million years ago the atmosphere must have been quite combustible and immense forest fires are probable. That prompted $O_2$ to swing back to more normal levels and $CO_2$ to rise.

It is difficult to say, of course, whether the thermostat really did kick in in such a way. Volcanoes, meteorites and other factors might also have played a role, but the point is that there are many chemical and biological feedbacks that all, with a certain probability, activate when the pendulum swings too far one way, thereby forcing it to swing back again. In any case, a billion years ago carbon and oxygen 'kept each other in sight' and they adjusted the pendulum through the relationships formed together and with other elements – in company, of course, with an assorted selection of minerals, perhaps most significantly with $CaCO_3$ and phosphorus.

As previously mentioned, we can observe the effect of current combustion by observing the $O_2$ concentration in the atmosphere, which is a mirror reflection of the Keeling Curve. Whereas $CO_2$ increases slowly and steadily, $O_2$ sinks in an almost stoichiometric relationship: 2 oxygen atoms per carbon atom. Nonetheless, the balance between carbon and oxygen in the atmosphere is so much in oxygen's favour that even a measurable decline in $O_2$ would give us no reason to worry that combustion was making a dent in atmospheric $O_2$ levels. Even the combustion of all known deposits of coal, oil and gas would consume a mere 1 per cent of atmospheric oxygen, so at least terrestrial organisms need not fear dyspnoea. That also means that previous $O_2$ fluctuations cannot be explained solely with reference to organic carbon.

Ocean life, in contrast, has significant reason to fear. Surface water in contact with the atmosphere normally exists in equilibrium with the atmosphere's $O_2$, and can even be supersaturated with $O_2$ in areas with high algae production. Down in the depths, on the other hand, light quickly vanishes, the speed depending on the amount of particles or brown humus material (dissolved organic carbon from land) present. At a given depth, called the compensation point, the quantity of light is so low that absorption of $CO_2$ and production of $O_2$ are in equilibrium. Below this level a net consumption of $O_2$ occurs and $CO_2$ builds up as the result of a steady, oxygen-devouring breakdown of what is produced in the upper layers, above the compensation point. This oceanic productive layer acts like a narrow, life-giving plane above the cold, dark and oxygen-imbibing depths. With the exception of the strange ecosystems around hot chimneys on the ocean floor, all life depends on the production taking place in the surface layers.

If the supply of organic material in the depths grows large enough, especially if there is little circulation between the upper layers and the depths, an oxygen deprivation in the depths results. This is not a crisis at a local level but, as previously mentioned, the ocean has undergone periods of massive oxygen depletion: partly due to consumption of organic material, partly due to chemical oxidation, partly due to reduced algae production (for example, on account of acidification), and partly due to warmer water. Warm water has a lower gas solubility than cold. The effect, in any case, is catastrophic for ocean life and dramatically alters the chemistry of the ocean's depths, both in terms of nitrogen and phosphorus, again impacting production. Historically there have been episodes of prodigious oxygen depletion in the ocean depths, the formation of unpleasant and toxic gases like $H_2S$, and massive extinction episodes, but broadly speaking these are connected to warm periods in the earth's history.

Therefore it is not without concern that we are now witnessing growing oxygen depletion in many deep ocean areas. Some of this is due to an increased supply of plant nutrients like nitrogen and phosphorus from human activity, some is probably due to warming, and some can also be attributed to increasing ocean acidification, though the latter does not seem to have done much so far. Higher surface temperatures, which can now be observed in most ocean areas, not only reduce gas solubility, but reduce the chance of mass water circulation since warm water is lighter than cold. Greater temperature differences, that is, will increase the chance that a warm 'lid' will occur on the surface. There is also a potential feedback here, since this will hinder the supply of nutrient-rich deep water. The result will be reduced algae production on the surface and less organic carbon 'raining' into the depths. No one can say if this is enough to counteract oxygen depletion. We can only hope.

The connection between carbon and oxygen is crucial to the earth's history, and an important if complex part of the earth's thermostat. Since oxygen enters into countless organic compounds, and oxidizes most of them, it, like carbon, is involved in most of what occurs. The oxidation of iron disulphide, for example, could have been a source of the atmosphere's historical loss of $O_2$. History, however, is complex enough even without us embroiling ourselves further in the countless promiscuous relationships that involve large swathes of the periodic table. Despite this, we cannot avoid carbon's two closest neighbours, namely nitrogen (N) and phosphorus (P). They have already, of course, been duly presented as actors in life itself. Their close connection to carbon within cells particularly ensures that they also play a decisive role as adjacent gears in the carbon cycle machinery.

## NITROGEN AND PHOSPHORUS: THE GREAT CIRCULATORY CHANGES

Even if the atmospheric $O_2$ sinks as a result of our global combustion processes, this is only a ripple on the surface of the atmosphere's prodigious $O_2$ reservoir. The atmosphere's nitrogen content is even more secure. Based on weight, there is 75 per cent more nitrogen in the atmosphere, while it occupies 78 per cent of the atmosphere by volume. Nothing we could conceivably do could make much of a dent in that, at least not within a reasonable time horizon.[55] Part of the reason for this is that nitrogen has linked to itself with a rock-solid triple bond, and therefore $N_2$ is what we call an *inert*, that is, a non-reactive gas. The air we breathe contains almost five times as much $N_2$ as $O_2$, but that is no problem. We exhale as much as we inhale. $N_2$ travels through the lungs, simply along for the ride. In the Scandinavian languages, nitrogen was previously called 'asphyxiating matter', *kvelstoff* in Norwegian, not because it is toxic, but because $N_2$ is useless for most organisms. Pure $N_2$, of course, quells life and fire, but it is entirely harmless as long as we have enough $O_2$.

The atmosphere's $N_2$ content has experienced ups and downs over a long time scale: the gas was not even present in the earth's very young atmosphere. Back then hydrogen (H) predominated in the atmosphere and nitrogen was represented in ammonium. Later $N_2$ and $CO_2$ were both pumped out of the earth's interior, and $N_2$ has been dominant for the last 4 billion years of earth's history. Around 2 billion years ago, the atmosphere was more than 95 per cent $N_2$, but has since been forced to yield a bit to $O_2$. Neither nitrogen nor oxygen has exhibited any lack of imagination when it comes to exercising mutual influence and forming mutual bonds, and their relationships, including those with other elements, are numerous. High on the list of partners are carbon and hydrogen. (Carbon usually has a couple of electrons free for longer-lasting as well as looser relationships.) When it comes to life itself, it is mainly in proteins that nitrogen makes itself known, with amino acids forming the building blocks within the unbelievably complex protein molecules. All amino acids are variations of the same theme: a carboxyl group (-COOH) and an amine group (-$NH_2$). The link between these groups is naturally carbon.

The nitrogen cycle could fill its own book, but in this case will have to play second fiddle, despite its overwhelming dominance in the atmosphere. The short version of the nitrogen cycle is that atmospheric $N_2$ is sequestered by some bacteria groups, among them blue-green bacteria, and this nitrogen fixation acts as the original springboard for almost all biologically available nitrogen – and thus for all additional life. Breaking $N_2$'s triple bond is no trivial task: it requires a decent amount of energy, the delicate enzyme nitrogenase and the absence of oxygen. Bacteria that perform this feat in the presence of $O_2$, for example on the surface of ocean, lake or soil, perform the job within specialized cells with oxygen-free interiors. Nitrogenase creates ammonia ($NH_3$), which is used by plants as the raw material for amino acids. At that point, nitrogen makes its way through the food chain, as Victor Summerhayes and Charles Elton described in their simple ecosystem on Bjørnøya. When members of the ecosystem die and decay, proteins and other materials containing nitrogen are broken down and can be absorbed again. Animals, of course, also need to rid themselves of nitrogenous waste products; humans do this via urine. After some trifling modifications, urine is once again accessible to plants.

Like urine, part of the reduced nitrogen in dead organisms and material will be greedily exploited by other bacteria groups that oxidize it in several steps, first into nitrite ($NO_2$) and then nitrate ($NO_3$) in a two-step denitrification process. Nitrate is the second key form of nitrogen that plants can utilize, but a portion of this will again be taken up by other bacteria groups that perform nitrate reduction – $NO_3$ via $NO_2$ and $N_2O$ back into $N_2$ – and the circle is closed. At this point the observant reader might ask what specific relevance this has for carbon, aside from the obvious remark that without nitrogen circulating through ecosystems there would be no carbon in circulation. Indeed, there would be no life at all.

The main connection between the nitrogen cog and the carbon cog is that we have significantly impacted the nitrogen cycle. There are two particular reasons for this. Every combustion process not only forms $CO_2$ but oxidizes $N_2$ to $NO_X$ (the generic term for NO and $NO_2$). Multiple processes can transform NO into $NO_2$, and $NO_2$ reacts with water to form nitric acid ($HNO_3$) in the atmosphere. This compound can travel large distances before it reaches the ground via precipitation, where it has two effects: it acidifies and it fertilizes. Both influence the carbon cycle: acidification increases weathering, and we have already seen what fertilization does with $CO_2$ (and phosphorus) – it performs the act it was meant for, namely, increasing plant growth on land and in the sea.

Our second key intrusion into the nitrogen cycle is the fertilization industry's Haber-Bosch process, in which ammonia is produced by electro-shocking hydrogen derived from methane and $N_2$ from the air. This is a key factor in the green revolution, but the flipside is that over-fertilization in agriculture yields an unintended and often problematic increase in the polluting of waterways and coastlines. The other side effect is large quantities of ammonia that vaporize from farmland and animal manure as it is spread. This ends up in the atmosphere and is distributed by the same winds as $NO_X$ and $HNO_3$, falls with the same precipitation, and produces the same consequences: acidification and fertilization.

In many areas, for example southern Norway, nearly equal amounts of oxidized and reduced nitrogen fall in concentrations many times higher than in pre-industrial times. In the short term, what we have done is to drain atmospherically stable $N_2$ and convert it into biologically active forms of nitrogen. We now convert more $N_2$ than nature itself, so this represents a fundamental interference. As previously mentioned, this fact will have no impact on atmospheric quantities of $N_2$, but will be felt in ecosystems as a large-scale fertilization experiment. For forest owners this is excellent news, as forests grow more than ever before and contribute to increased $CO_2$ uptake on land. More nitrogen also increases ocean production and it is probably our intervention in the nitrogen cycle that is part of the explanation for why the Keeling Curve is not any steeper.

Since $NO_X$ also plays a significant role in methane breakdown, here nitrogen also comes out on the plus side of our climate balance sheet. Nothing is so bad that it is not good for something. On the other side, we can again cite the pessimistic variant: nothing is so good that it is not bad for something. The remarkable increase in the greenhouse gas $N_2O$ is also the result of the same interventions and cancels out some of the positive $CO_2$ effect. Increased oceanic algae growth, furthermore, not only contributes to increased carbon binding, but can contribute to oxygen depletion in the depths.

Regarding the ocean's depths, the same thing happens to the nitrogen when the ocean suffers oxygen depletion as we observe in lakes or ocean areas with oxygen-free depths. In part $NO_3$ is converted to $NH_3$, in part denitrification speeds up (nitrate-reducing bacteria also thrive in the water), and nitrogen is degassed back into the atmosphere as $N_2$. The less oxygen, the better it is for denitrification, and eventually the ocean will suffer a critical nitrogen deficiency. Algae production will slow and more $CO_2$ will remain in the atmosphere. With its complex cycle, nitrogen indeed has a finger in most of what happens and this is truly one of the great cogs in the machine.

What about phosphorus? In its modest way, phosphorus does not even make a natural loop through the atmosphere, it figures low on the list of bodily elements and even lower on the list of earth elements. Phosphorus is one of the most humble, but nonetheless indispensable of neighbours, and we have already discussed why it comprises a key part of the biological collective.[56] Phosphorus's scarcity is exactly what makes it so valuable, since value is obviously a result of the balance between supply and demand. Quantitatively speaking, demand is perhaps not so great, but the supply is even less, which makes phosphorus one of the most prized elements in the biosphere.

Compared to carbon and nitrogen, furthermore, phosphorus can hardly be said to have a proper cycle. It weathers from mountains, often from apatite (for example, hydroxylapatite, $Ca_5(PO_4)_3OH$), and often with the help of plant roots. Of course, plants are notoriously interested in phosphorus themselves, so one might assume that plant roots would be hesitant to surrender what phosphorus they have wrestled from the ground so laboriously. That assumption is entirely correct, but there is still a little phosphorus left over. Trees can take up phosphorus only in the form of phosphate ($PO^{4-}$), and this means that small weathered particles with intact apatite can be transported to rivers and ocean where weathering can continue. As such, the phosphorus cycle is a fairly predictable affair.

Much of this weathered phosphorus ends up in the ocean where, like carbon and nitrogen, and usually in conjunction with these, it circulates through the food chain before eventually disappearing into the deep layers and coming to rest in the sediment. It is estimated that every phosphorus atom is recycled fifty times on average before it vanishes from the system. If there is an abundance of iron present, it will seize some of this phosphorus and transport it to the depths, beyond the reach of producers. Here phosphorus will normally be kept out of circulation for long, long periods of time, until one day it is again thrust to the surface by mountain chain formation or volcanoes. Since phosphorus does not naturally exist as a gas, it will fall along with ash to the earth, but most follows the laborious process of mountain formation and erosion, and we are talking here about a truly slow cycle, one that lasts millions of years. As previously mentioned, it is quite possible that one of the reasons the globe cooled after the great warm period 55 million years ago was the creation of the Himalayas at around that time. While not exactly an instantaneous process, continental plate activity can lift enormous land masses by many kilometres. Steep mountain walls and gushing rivers provided a massive weathering and transport of phosphorus to the ocean, increased algae growth, and a powerful enough absorption of atmospheric $CO_2$ to drive temperatures down.

The only factors that can disturb this sedate and uncomplicated cycle are those that govern the Urey Reaction. Nonetheless, this has been enough to significantly impact the slow thermostat, and the simple result of more accessible phosphorus has always been increased plant growth, more oxygen and, primarily, less $CO_2$. Since nitrogen and phosphorus are so intimately connected in protein synthesis, there will always be a need for a sufficient quantity of both. It does not do plants much good (nor indeed ourselves) to have unlimited access to phosphorus if there is a shortage of nitrogen, or vice versa.

Despite phosphorus's rather prosaic cycle and limited quantities (or perhaps because of them), the phosphorus cycle is the one we have truly upended. There is an almost insatiable need for phosphorus because, unlike nitrogen, it cannot be retrieved from an inexhaustible atmospheric reservoir. For a time we recovered phosphorus as fertilizer from guano, that is, bird dung, for example from old deposits beneath nesting cliffs. To some extent this is still done, but there are limits to how long supplies will last in a world with a steadily increasing demand. Today's demand for phosphorus is supplied by phosphate mines, though there are not many places in the world with commercial deposits. Five countries control more than 90 per cent of the existing resources: Morocco (with the annexed Western Sahara), China, South Africa, the USA and Jordan. Of these countries, Morocco is clearly at the top: if I were to buy mining shares, it would be for Morocco's phosphate mines. There is nothing that can replace phosphorus, and in one or two hundred years we could face a chronic shortage of this essential element, which today we regard as troublesome, whether as run-off from farmers' fields or untreated sewage, resulting in massive algae growth and poor water quality. 'Peak phosphorus' is approaching and will be a much greater challenge than 'peak oil', but to the human brain a hundred years from now still seems far away.

We have altered all the globe's great cycles, but few so fundamentally as the phosphorus cycle. Whereas our annual contribution of atmospheric carbon is less than 5 per cent of the ecosystems' contribution (our yearly emissions are 9,000 million tons against the ecosystems' 210,000 million tons), our fixation of atmospheric $N_2$ exceeds the ecosystems' fixation (150 against 130 million tons annually). Yet when it comes to phosphorus, we have contributed to a 400 per cent increase in the quantity of phosphorus set in circulation (12 million tons from phosphate mines against 3 million tons annually by 'natural' weathering processes).

The world is complicated and researchers strive for insights that can lift the veil. On the flip side, research has shown us how truly complicated everything is. Therefore, the reductionist method is science's primary approach, isolating individual phenomena and details, and attempting to understand a single piece of the puzzle before trying to piece the whole thing together. At the heart of all these complex systems and questions are simple natural laws, and physicists love to point out that 'everything is physics', usually described in mathematical terms. There is an element of truth to that, and the basis for the carbon cycle and $CO_2$'s effects on the environment is directly anchored in physics – and chemistry. However, we do not have enough information to fully explain the carbon cycle. As such, it is not without reason that no matter how much data is fed into climate models, there will always be a considerable amount of uncertainty, although this uncertainty is more evident in the details than in the general picture. The most significant of all these uncertain factors is the different ways we humans will respond to carbon. A 'theory of everything' will always be beyond our reach.

When I am teaching, after making my way through lectures with pedagogically appropriate trophic pyramids, food chains or food webs (which naturally are a little closer to reality), I show a wonderful figure entitled 'A Simplified Food-web of the North Atlantic'.[57] This is an apparently chaotic tangle of lines that connect together the various actors in the North Atlantic's food web, thereby underscoring ecology's dilemma: everything is connected. This, though, is actually a 'simplified' version, with algae appearing as a single group, copepods as another, and so on. To truly rub it in, the figure also shows a slice of the metabolic network as it occurs inside organisms and cells. This picture could again be divided into carbon, nitrogen, phosphorus and so on.

This gives the impression that it is complete chaos out there, that every attempt to distinguish patterns and predict effects is doomed to failure. How is it possible to say anything, for example, about the carbon cycle in ecosystems with this kind of disarray? It is possible, though, to offer some guidance since, for example, some arrows are shown as thicker and more important, while some species are key and others more marginal.

Our tour through the carbon cycle has probably given much of the same impression. What we have is a conglomeration of causes and effects in time and space, chemistry and biology. That impression is correct: we do not need to access the Beilstein database of 10 million organic compounds in order to remind ourselves that nature offers plenty of complexity. Just understanding the humus molecule's structure, genesis and fate is a task far exceeding a life's work. We have only followed the roughest arrows, but that is often enough. If we know the principles and the most important actors, it is not necessary to follow all the intricate paths. Of course, this still assumes we have enough knowledge of the processes, arrows and boxes to enable us to pick out the most important species or important reactions and the specific roles they play. An exhaustive knowledge of all metabolic functions, for example, is not required to understand that an exclusively carbohydrate diet is not particularly healthy. Similarly we do not need to know all the details of the carbon cycle to conclude that we are in the process of impacting it dramatically.

It is the human-induced 4 per cent of emissions to the atmosphere that has created Keeling's Curve: if all our emissions had remained in the atmosphere, the curve would have been steeper. Positive or self-reinforcing feedbacks are a set of different mechanisms that strengthen warming effects. The key terms are albedo, thawing of the permafrost, the drying out of forests and soil, and ocean acidification. The essence is that positive here means negative: in technical terms we are talking about positive or self-reinforcing feedbacks that, in this context, have significantly negative consequences. The thermostatic effect here implies that when effects become extreme, counter-reactions will occur. The problem is that most of these are slow and of limited use in the situation in which we now find ourselves. There is little comfort to be found in the old saying that 'everything will be forgotten in a hundred years' when you are left blushing in embarrassment for what you have done. We need to cut our emissions *now*, before we set in motion a chain of uncontrollable feedbacks. Maybe we can install our own thermostats? Dump iron into the world's oceans to increase production? Hardly. Dump more nitrogen to fertilize the forests? Hardly. Stop deforestation? Absolutely. The one critical insight, however, is this: we pump out too much $CO_2$. The only logical and uncontroversial conclusion here is to reduce emissions, but how?

# PART III

# THE FOOTPRINT

The earth and life itself have been through a number of trials; the pendulum has oscillated several times between greenhouse and snowball. Impressively enough, the pendulum has managed to swing back again each time the situation became critical. Life has persisted and continued to develop, even as it helped to stabilize the pendular situation. Biological processes have proven decisive in stopping the pendulum's most extreme swings, those that could well have locked the globe in rigor mortis. A series of different feedbacks have prevented the earth from spiralling back to a lifeless and molten Venus-like state, in thermodynamic equilibrium now because respiration has ceased, or, alternatively, gripped by an eternal freeze. Throughout the planet's history, such involuntary control mechanisms, such thermostats, if you will, have been physicochemical and biological in nature, although in the last 500,000 years the biological mechanisms have proven more significant. Biological mechanisms, furthermore, are the most significant on time scales relevant to us.

The pendulum may have swung back, but its oscillations have not been without cost. We can trace at least five mass extinction periods in the earth's history, all tied to climatic extremes. Altogether 98 per cent of all species that have existed on earth have died out, many of them because it was simply their time as they were outcompeted by species who evolved better strategies to meet life's challenges, but others due to greater or lesser catastrophes. Yet are these really catastrophes? Death for some means life for others; some species end up on history's scrapheap, others stand poised to take over. On behalf of all mammals, we can probably thank the meteorite 65 million years ago for our success. The drought that once forced Africa's forests into retreat, thereby prompting our ancestors down from the trees and out onto the savannah as the first step to conquering the world, was also largely the result of a changed climate.[1] In particular, we can thank blue-green algae and their numerous relations for providing the atmosphere with oxygen. Although this was great for us, it was a real crisis for ancient organisms that had evolved to live in an oxygen-free environment.

## PARADISE LOST II

During the last 800,000 years the atmosphere's $CO_2$ level has vacillated between 180 and 280 ppm, the minimum coinciding with ice ages and the maximum with warm periods. Current levels are around 440 ppm, and surely nothing can prevent us from reaching 500 ppm. What the result will be no one knows. The Anthropocene is characterized not only by human interference in the great cycles, but by our numerous and notable interferences into the world's ecosystems, in terms of both land usage and effects on animal and plant populations. For this reason, the Anthropocene is also characterized by mass extinctions. Until now this has been due to hunting and habitat destruction, but eventually climate change will likely prove to be the greatest threat. We are heading, in fact, for a sixth mass extinction and it could mean the end of the world as we know it. Presumably, *Homo sapiens* as a species will survive. If there is anything at which we are experts, it is weathering the most inhospitable conditions, but it could very well mean untold suffering and deprivation. From this perspective, it is a poor consolation that we seem to be back to the 200,000 year norm, although for the ocean we might say 500,000 years. *Homo sapiens* as a species is only about 100,000 years old, so we are talking about a long perspective here, even if it seems short in a geological and evolutionary context. The point is that, although life as such always seems to continue, many of the species we now know and treasure, and countless others as yet unknown, will be lost forever. The world will never *completely* go back to what it was.

In autumn 2014 the WWF (World Wide Fund for Nature) published a meta-analysis, a compilation of notable literature concerning the development of world animal populations: mammals, birds, reptiles, amphibians and fish.[2] On average, the last forty years have seen animal populations cut in half. A halved population does not imply that a particular species will disappear, but it is tempting in this context to recall the passenger pigeon. Martha the passenger pigeon died on 1 September 1914 in the Cincinnati Zoo, the last of her kind, the last in a species that figured just a century earlier as the world's most numerous bird. There were once somewhere between 5 and 8 billion passenger pigeons in North America; the first settlers told stories about flocks so vast they blackened the sky. Yet these birds were hunted mercilessly long after it became clear that the population was in catastrophic decline. The last wild passenger pigeon was shot in Ohio in 1900 and Martha, the ultimate representative of her species, followed the auk in 1914. Hunting spelled the end of Martha and her predecessors, not climate change, and the world continued on its course without passenger pigeons and auks, as it will also do in the wake of many others. The extinction of the panda, if it came to that, would not impact the globe or the climate, but it would still represent a loss, even if (from a human perspective) the clear or directly beneficial value is difficult to pinpoint.[3]

Many species, of course, can disappear without prompting fundamental changes to ecosystems and certainly not to the climate. The point remains, though, that population decline and species loss can happen at a furious pace. Species that have evolved through millions of years can vanish in the course of a few decades, and climate change will escalate this process. In fact, it is estimated that between 30 and 40 per cent of the world's mammals and birds could face extinction within the foreseeable future as a result of climate change. No one, of course, knows for sure: it might be more like 10 or 20 per cent, but even those numbers are dramatic. On the other hand, if entire groups, such as *Emiliania huxley*, or entire ecosystems, such as the Amazon, were to decline or disappear, the effects will be immense, and it will be a dearly bought confirmation that there indeed exists a mutual interaction between climate and nature. We can also hold out hope that species and ecosystems will adjust to the new conditions. Given their rapid evolution rate, microorganisms can persist for a while, though changes are also occurring at a rate that outpaces evolution itself. Even PETM, not to mention the Cambrian explosion, happened in slow motion compared to what is now taking place. That previous catastrophes were not worse is due to the fact that many species were simply able to keep up.

When it comes to biological production, the world's remaining spaces are also declining. An estimated 80 per cent of the globe's productive land area has to some extent been affected by human activity, and 36 per cent of the globe's total land area is dominated by human activity. That fact, together with increased harvesting and eventual climate effects, will also take a toll on the planet's diversity. As we know, the world's rainforests have likewise been reduced to half in little over a hundred years. In the midst of all this, the species with the greatest flexibility, the greatest recourse to various forms of aid, and which also happens to populate every nook and cranny of the globe, is human beings. From this perspective it would seem we would also have the greatest chance of riding out the storm – if not for the fact that we are completely dependent on other species, on the products and services that ecosystems provide, for everything from oxygen production to food. It is quite a paradox, then, that a species ingenious enough to dominate the globe so totally that it has an entire geological epoch named after it must also recognize that the so-called Anthropocene is largely characterized by negative developments. This would be a form of herostratic fame, if others were to come along after us and evaluate this geological epoch. We should be a *Homo sapiens* who earned the name not only for ingenuity, but for wisdom. Yet for all our knowledge, we have perhaps overreached ourselves, purchased shoes that were too dear, paid for in instalments. Or is the problem not price but size?

## SHOE SIZE

Once a year the organization Global Footprint Network celebrates something called Earth Overshoot Day.[4] In 2016 it fell on 8 August. As the name implies, it is the day by which we have consumed the earth's production capacity for that year; the rest of the year we are, strictly speaking, depleting future resources, for example by lowering groundwater levels, precipitating soil erosion and exploiting what remains of untouched natural spaces. Our carbon footprint is also taken into account: in fact it constitutes 54 per cent of the index (a portion that is steadily growing). From this vantage point it is rather like taking out new consumer loans to cover current expenditures. Based on available statistics for potential biological production, which takes the form of agricultural products, grazing land, fisheries, imports and exports, as well as the consumption of the same, the index also calculates for 150 countries whether they appear on the surplus or deficit side.

Whereas in 1960 most countries could celebrate a surplus, the tables have largely turned for most of the countries examined. If nations like Brazil still find themselves on the plus side, with a bioproduction capacity exceeding consumption, it is because the consumption rate there is still relatively low. Of course, Brazil's actual capacity has been halved since 1960, not owing to its own inhabitants, but rather to pressure from the outside world for products such as soy, palm oil, beef and timber.

As for the carbon footprint, it is calculated by considering $CO_2$ emissions relative to the capacity of ecosystems and agricultural land areas to absorb $CO_2$. If green regions capable of conducting photosynthesis shrink while emissions rise, logic dictates the fraction will shrink (the footprint will grow). This figure, though, is not terribly precise, since the numbers in the equation cannot be calculated exactly and, furthermore, will be affected by factors such as increased nitrogen deposition (which increases carbon absorption), climate changes (which can go both ways) and ocean acidification (which decreases carbon absorption). Nonetheless, the figure is precise enough to reveal trends and magnitudes and to confirm that our present course is not good.

What is the largest, cumulative footprint we can leave behind and still have some reasonable hope that the carbon cycle will not run amok? IPCC provides an answer to this question: if we want to maintain a 66 per cent chance of staying within the infamous 2°C, we can at most emit 800 billion tons of carbon (or 800 Gt C, which amounts to 2,900 Gt $CO_2$).[5] That number certainly sounds exact and it also sounds huge.

In fact, both impressions are wrong. That upper level is only a best estimate and a 2°C increase is no precise or magical boundary. It is only put out there as a postulation regarding when the thermostat (the short-term one) may clock out and the positive feedbacks take control. Furthermore, an average of two degrees will mean enormous local and temporal variation, not to mention extreme weather events, and many climate scientists believe that the two-degree target is actually too high. Nonetheless, it is important to establish some goal and two degrees is the best we have, even if it currently appears more and more utopian.

Yet is 800 Gt C actually a lot? Taken by itself, it is undoubtedly huge, but it does not sound quite so impressive when we realize that since our 'reference period' of 1861–80 we have emitted 520 Gt C. In other words, we have only 280 Gt to go. As mentioned above, other people have placed the atmosphere's residual capacity even lower: David Archer, someone who truly knows his carbon cycle, estimates 700 Gt, whereas Bill McKibben and his organization 350.org have pinpointed a 565 Gt residual capacity based on climate researcher James Hansen's calculations. The lower we go, the better, but let us be done and settle for IPCC's calculations. We therefore have an allowance of 280 Gt C to last us from 2011 until 2100.

At present, human-induced emissions are about 10 Gt C annually. This is 61 per cent higher than the Kyoto Protocol's reference year, 1990. China is responsible for 28 per cent of these emissions, followed by the USA (14 per cent), the EU (10 per cent) and India (7 per cent). The trend is also increasing for every region, with the exception of the EU, which exhibits a slightly falling trend. Coal is the true miscreant here and contributes 43 per cent of emissions, followed by oil (33 per cent), gas (18 per cent) and cement production (5.5 per cent). Notice, however, that these figures are only for $CO_2$/carbon. Methane and other warming sources, like nitrous oxide and soot particles, are not included.

The course has been charted and investments made for extracting fossil fuel reserves totalling 762 Gt C, nearly three times the maximum residual capacity. If we take today's emissions as our basis, 10 Gt C per year, we will have spent our allowance in 28 years. Of course, that supposes that our emission level remains stable. Since ecosystems manage about half our emissions, we can actually double our emission level based on the optimistic and probably unrealistic expectation that ecosystems will continue to perform this function. In that case, we have 56 years left.

At the time of writing, the world's population totals in excess of 7,500,000,000 individuals. If everyone in the world were to emit as much $CO_2$ as a typical Norwegian (about 9 tons $CO_2$ annually), emissions would rise to more than 65 Gt $CO_2$ (18 Gt C) per year, which means the 'magic limit' will be exceeded in only fifteen years. Every day the world's population increases by around 210,000 people and in just ten years it will have passed 8 billion. If a growing portion of these 8 billion people should creep up toward Norwegian emission levels, there is only one solution. We must all descend, not to the basement, but to less than a quarter of today's emissions – and that as quickly as possible.[6]

Depending on the scenario one accepts, this means that between about two-thirds and three-quarters of the planet's remaining fossil reserves should stay untouched: coal certainly, but also oil. The question here is whether this is realistic at all, since investments in existing resources have a market value of $27,000 billion. There are also powerful forces that desire to avoid at any cost the fossil account being barred from withdrawals. Of course, not everyone formulates this idea as explicitly as Rex Tillerson, former CEO of ExxonMobil, the world's largest non-state-owned oil company. What are known in the trade as 'stranded assets', that is, unused investments, simply do not exist, Tillerson commented, turning around and investing $40 billion in oil exploration in some of the world's most vulnerable regions, from the heart of the Amazon to the western Antarctic.[7] Everything invested in fossil resources, now or in the future, must re-emerge. Even if few express themselves as clearly as ExxonMobil's erstwhile head (and currently the U.S. Secretary of State), it is obviously a premise of operation for many who own fossil resources, whether these are private individuals or countries. The same countries that meet for climate talks with furrowed brows are continuing to subsidize exploration for fossil reserves through granting tax breaks to exploration agencies. According to the Overseas Development Institute, the world's richest countries annually spend around $90 billion on tax subsidies for exploration ventures looking to extend the current fossil period.

If we accept the IPCC's premise, along with that of most climate researchers, that there is indeed an upper limit and, therefore, a residual capacity of 800 Gt C, that means that whatever we emit now cannot be emitted later. The longer we wait to reduce our emissions, the more we will have to cut later, leading to greater costs both economically and socially. If we begin today, we can make do with a 2 per cent annual reduction; if we wait until 2040, with business as usual until then, we will be forced to cut 35 per cent per year. That strategy does not seem entirely smart; sooner or later we will be forced to adjust to a fossil-free existence, but part of the problem is that a significant number of people, companies and countries will be reluctant to accept the idea of invested capital remaining in the ground (the general idea among many fossil nations and companies seems to be that oil should be extracted as long as it is still worth something).

The much celebrated green shift is under way, but it is not clear how quickly it will reduce our interference in the carbon cycle, and we also cannot simply wait for the market's invisible hand to sort matters out. When David Keeling warned about the potential effects of our supplying the atmosphere with additional $CO_2$, the annual emission rate was 4 Gt C. When the Brundtland Report concluded in 1987 that there exists a critical 'need to change course', emissions had risen to more than 7 Gt. We are now approaching 10 Gt yearly, 90 per cent of which comes from fossil energy, 10 per cent from deforestation and other land interventions. The carbon cycle's ongoing reaction will be crucial to our chances of remaining beneath the residual capacity's quota. If our carbon emissions (which we can actually do something about) set in motion cyclic feedbacks (which we can do nothing about), our position is not good. Human history has to a great extent revolved around gaining control of natural forces and taming nature to our advantage. We have largely succeeded, with the exception of what exceeds anyone's control: the weather. If we actually manage to impact the weather, as a result of what we are doing to the climate, it paradoxically represents a loss of control.

## HOW BAD ARE BANANAS?

For most products and activities, $CO_2$ totally dominates the gas budget. Yet often nitrous oxide and methane are more subtly involved, particularly when it comes to agricultural products, which are responsible for up to a quarter of all global emissions. In order to include their contribution, it is expedient to use $CO_2$ equivalents ($CO_2$e), which weight the contribution from each gas according to their relative warming potential, a strategy adopted by Mike Berners-Lee in *How Bad Are Bananas?* (2011), an entertaining and informative look at human consumption, covering everything from emails to bananas to airplane trips.[8] How much $CO_2$e does a human life contain and how much can or should it contain?

It cannot be repeated too often that a carbon cycle gone awry is the sum of 7 billion people's relationship to beef, cars, living space and everything else. One reoccurring argument is that our consumption is not at fault, but rather the combustion of coal, oil and gas. This is certainly true, but for whom is that coal, oil and gas burning? In whose name is the rainforest burned in order to make way for palm oil plantations, soy acreage or grazing land? Keep these questions in mind as we take a look at beef and bananas, as well as other things we combust directly or indirectly. How much $CO_2e$ does an email cost, or a PC in sleep mode, or a bread loaf, or 10 miles driven in an SUV compared with 10 miles in a compact car, or a long weekend trip, or a family holiday to Thailand?

It is not the primary intent of these calculations to produce a guilty conscience, although some guilt might be in order, but rather to help set our life in perspective. Recently I have visited the debate regarding which is the more energy-friendly alternative: drying hands with a paper towel or using an electric hand dryer. Such a discussion, of course, quickly becomes meaningless if you are chatting with your family in a Mediterranean hotel: a Mediterranean holiday is not necessarily morally reprehensible, but in terms of pure carbon, you would have to stand before an electric hand dryer for a good few years before it began to compare with a few hours spent in the sky. And that is exactly how we ought to think: everyone has an annual $CO_2$ quota that will maintain us within responsible limits. You can decide for yourself how you will realize your quota, but that means you will also have to choose.

Exactly how bad are bananas? As it turns out, not so bad, although they do come from abroad and obviously exact their morsel of $CO_2e$. As such, we can keep eating bananas with a relatively good conscience. Steak, on the other hand, has a bad reputation in environmental circles, but just how bad is steak? Is it better or worse than a burger, and does the steak or burger I am eating play a role in the great balance sheet? Does it play a role, furthermore, whether I eat tomatoes grown in my own country or ones that come from Spain instead? Starting with the burger, a standard burger, including the efforts of agriculture, fertilizer, nitrous oxide and methane emissions, carries a cost of 2.5 kg $CO_2e$. Eating a burger a day, which is not a recommended diet for a variety of reasons, will total 910 kg $CO_2e$ over the course of a year, the equivalent of driving about 2,500 km in an average car.

Since it is only meat, steak emerges quite a bit better than burgers. Burgers, after all, accumulate additional costs with their extra ingredients. An average steak requires 2 kg $CO_2e$, the equivalent of about 40 km in an average car. The deciding factor in this context, furthermore, is not the distance you drive to the store in order to buy the steak. It is not even whether the meat itself has travelled a short or long distance to get there. The difference between local and imported beef contributes only a few hundred metres to those 40 km. Steak's formidable footprint is due, in part, to the costs of feed production, tractors and other $CO_2$-generating processes, and in part to the fact that cattle are such significant methane producers. Because of nitrogenous fertilizers, agriculture is also a substantial nitrous oxide emitter. It makes a difference, therefore, whether the animals involved graze on grass or are fed soy, especially soy imported from areas that were once rainforests.

What about eggs? Believe it or not, a carton of twelve eggs has an even larger footprint than steak, 3.6 kg $CO_2e$. How is that even possible? The answer has to do with the intensive agriculture required and also because a carton of eggs represents a solid source of nitrous oxide. In total, $CO_2$ comprises 41 per cent of the equation here, whereas nitrous oxide accounts for 45 per cent due to chicken feed. Methane also has a small portion to contribute, whereas the actual transportation amounts to only a marginal 1 per cent. The origin of tomatoes is a perpetual topic of discussion in green circles, since an imported tomato from Spain is more climate-friendly than one grown in north European countries, where the crop is cultivated in a climate not meant for it and requires intensely powered greenhouses. As such, tomatoes are bad news unless they are grown naturally outside in the summer heat (0.4 kg $CO_2e$ per kg): in winter, greenhouse-cultivated cherry tomatoes can register upwards of 50 kg $CO_2e$ per kg. Tomatoes may be healthy, but they are certainly not healthy for the climate.

What about wine with dinner? In this case, it is not the carbon footprint that suggests limiting yourself to one glass, especially not in comparison with steak. The equivalent of one bottle of wine, but packaged in non-imported cardboard, constitutes perhaps 400 g $CO_2e$ per glass. The Scandinavian countries, of course, do not produce much wine, and so there an imported bottle is just over 1 kg $CO_2e$, depending on whether it is transported by truck (worse) or boat (better).

Exercise also increases combustion and that means more $CO_2$, but you should be able to exercise with a good conscience; there are limits to how carbon-frugal one should get. (Instead, consider whether you are exercising because you feel you have eaten too much steak.) On the other hand, if you enjoy a hot bath afterwards, be aware that you are using up to 3 kg $CO_2e$. This is not entirely true in countries that create a high proportion of their energy by renewable technologies, and their widespread use in replacing coal-based electricity would have a marked impact on the global energy market.

Speaking of the globe, what does it cost to see it? Here is where we reach the truly large figures. If you are taking a short trip and are content to drive a small, petrol-powered car at moderate speeds, you are looking at 560 g $CO_2e$ per kilometre. However, if you want to leave the neighborhood and drive 1,000 km, you are looking at 56 kg $CO_2e$. If you take that same road trip at high speed in a large SUV, it is far worse: 400 kg $CO_2e$. And that is not including the costs of producing the car, itself no simple calculation considering all the components that must be traced back to their source. Aluminium necessitates the extraction and smelting of bauxite, and must be shipped and converted to bodywork and wheels. A car is also jam-packed with all kinds of polymers, electronics and batteries. An electric car, incidentally, is by no means carbon neutral, just in case anyone was wondering. There are costs here as well, particularly when it comes to the battery, which has to go somewhere when the car's life is over. On the global energy market, electricity is not carbon neutral, so an electric car will also carry a footprint, even if it is quite a bit smaller than diesel or petrol cars. It is when you hop onto a plane, however, that the numbers really take off: a couple of long flights a year can well exceed your SUV's annual total.

If you choose to stay home instead, and perhaps save money to build a house, that does not necessarily mean much gain carbon-wise. Building an 'average' new house costs an estimated 80 tons $CO_2e$. In this context, it is the cement associated with the house that drives up the price: the cement industry is one of our largest emissions contributors, though wood is not entirely absent from the equation. Clearing a hectare of forest equates to 500 tons $CO_2e$, but because a house will continue to store carbon, the calculation here acquires a time component. Yet the greatest factors driving deforestation continue to be items we consume, including steak, palm oil and soy, as well as other things with a space requirement.

And what about having children? Here it is not the energy costs of conceiving a child we are talking about, but the cumulative consumption that a standard Western child will expect to maintain throughout the course of his or her life. In this case, we end up with something like 700 tons $CO_2e$. That isn't an argument for remaining childless, but it is worth reflecting on how many children we should have and how we can motivate them to adopt a low carbon lifestyle, which is easier said than done, as I myself can attest.

Is there anything we *can* do in good conscience besides lie on our backs in the grass and watch the clouds float by, or don woollen sweaters and whittle by the hearth? Of course! When it comes to quality of life, the things that rank highest on the list are carbon neutral: love, family, friends, nature and, yes, even wine. On the other hand, much that causes stress and discomfort in our existence figures high in the $CO_2e$ category. There are exceptions, of course, such as exotic trips. Many people are fond of steak and cars, many people like to shop. The point here is not an ascetic life of denial, but rather that halving the number of trips, halving the amount of steak, halving the quantity of new *things* will also halve our carbon footprint, probably without impacting too much on life's pleasure. It might even increase our enjoyment and bring with it a dash of good conscience.

Even driving looks very different if, for example, there are four seated in the car instead of one (as a simple calculation will show), if the car is small instead of large, and if your pace is smooth and moderate instead of aggressive. In most areas today's consumption is many times higher than it was in the 1960s, a time when many people thought we had reached an impressive level of material prosperity. It is entirely possible we could cut our personal $CO_2e$ footprint in half without feeling notably deprived. An increased sense of freedom and improved quality of life might even be a bonus. As we know, the best things in life are free. In a comprehensive study regarding which pleasure-giving activities Norwegians rank highest, shopping and air travel did not top the list, but rather sex – which is not, perhaps, all that surprising.

Nonetheless, we all participate in the treadmill race of life and have an unfortunate tendency to measure ourselves against our neighbour's consumption – at least if that neighbour has a little more than we do. A natural place to change this attitude is to consider our personal $CO_2e$ quota. Instead of a legislated $CO_2e$ quota we can envision a moral one. Within that framework, we could decide whether to spend the day working out in the gym or hopping on a plane for a day trip shopping in Paris.

Those of us in wealthy countries average nine tons of $CO_2$ yearly, but we also consume goods and services, especially agricultural ones, that generate $CH_4$ and $N_2O$, both of which can be converted to $CO_2$ equivalents. For simplicity's sake, let us therefore say that our average $CO_2e$ is 10 tons annually (that estimate is probably too high, but we have to start somewhere). I am definitely not an environmental hero, yet I am increasingly aware of my responsibility. In February 2017 I was invited to receive an award related to my early works on stoichiometry at a scientific meeting in Hawaii. After some hesitation, I elected to save some $CO_2e$, as it would have put me way above my annual quota. I am not arguing here against conferences and meetings in general, but simply saying that we should perhaps consider their number and purpose, deciding whether a conference is more worthy than a holiday in Thailand. I almost always decline day meetings that require air travel, where it is just as easy to accomplish the task by Skype, telephone or email. A personal meeting is sometimes necessary, but this is rarer than you might think: not only do you save on carbon, but on time and stress.

Many of my colleagues appear to spend more time travelling than in the office. Some may have a personal $CO_2e$ footprint of about 100 tons a year, yet they are sincerely concerned about the climate. These are clever people who have certainly given the matter some reflection, but probably rationalize that the positive goals of travel outweigh $CO_2e$ concerns. It is not just my colleagues, however, for everyone flies. Airplanes are full, airports are expanding; throw a little wood-based bioethanol into the tank and one might even imagine they were taking a green trip.

Norway, of course, has declared its intention to reduce its emissions, and yet increasing demand has brought growing numbers of flights and expanding airports. Norwegian government employees on official business fly more than 800 million km a year, the equivalent of 15,300 trips around the world. There is nothing to indicate the private sector is flying any less, and this does not include any long weekends in Rome or exotic safaris. Few things illustrate the carbon dilemma more than air travel. Generally speaking, flights are not something we *must* take; on the contrary, communication technology ought to have prompted a decrease in air traffic. Yet air traffic emissions have seen among the largest increases of any sector and it is estimated that air traffic will continue to increase by 2.3 per cent a year until 2040. By clever accounting, only internal aeroplane traffic is noted in Norway's national carbon budget, and so the country's overall contribution appears modest, although it still totalled 53 million tons $CO_2e$ in 2016. That means, for instance, that only a fraction of the government employees' 15,300 trips around the equator are include here. The same logic is applied to calculating Norway's oil and gas sector. If the official total of $CO_2e$ that originates from the country's oil and gas production is expanded to include what is combusted abroad, Norway's contribution to global emissions increases from a few thousandths to 1.5 per cent – the same as Britain's portion of global emissions.

MY FATHER grew up on Hessa (hence my surname), an island on the west coast of Norway. He lived on a small farm where his family fished and kept livestock, which was then typical along the whole coast: a few cows, a dozen sheep, some pigs and chickens, as well as a horse to pull the plough and the hay wagon. There was also a boat for catching cod. It was a society that was almost carbon neutral. They used electricity for lighting, of course, and the Norwegian energy market was almost entirely electrical. My grandfather imported the first engine to the island in the form of a motorboat that could transport milk from the island to the mainland. Otherwise, it was an almost entirely natural economy where practically everything was self-produced: 90 per cent of dinners featured fish, the rest were meat or eggs from the farm's livestock. The sheep gave wool. Yet in my father's lifetime we have burned almost exactly half of that maximum quota of 800 Gt C, something it took several hundred million years to wring from the atmosphere. These decades mark a unique epoch in the planet's history. Over a couple of generations the world has undergone what has probably been an unprecedented transition from a natural economy to a petroleum society; we may well emerge on the far side as dry petroholics. Our way forward, however, does *not* lead back to a society with oil lamps and horsedrawn carriages.

In terms of the carbon budget, steaks, cars and living spaces are not beneficial, and clearly neither is flying – or tomatoes, or baths, or leisure travel. Must we then reinvent the wheel, which in practice means returning to my father's childhood society or to the 1960s? Would that enable us to live without guilt? Another possibility, of course, is simply to kick back and wait for the 'green shift' to come along and fix everything. A car today, for example, emits only half of what one did back in the 1960s, so why turn back the clock? Nor did we then have electric cars. Germany, once Europe's industrial powerhouse and renowned for its coal-black Ruhr region, is now largely powered by renewable energy. Indeed, China is also investing heavily in this area. We value carbon-neutral activities like love and friendship above all, but we cannot live on air and love alone.

## PETROHOLISM'S WITHDRAWAL

Germany's renewable sector is world leading and has increased its electrical supply from 6 per cent in 2000 to 30 per cent in 2014. Germany has also succeeded in reducing its emissions from 10.9 to 9.1 tons $CO_2$ per year per inhabitant in that same period, even if renewables still comprise only a small portion of its energy usage. China, in contrast, has climbed from 2.6 to 6.2 tons $CO_2$ per year per inhabitant in that same period, and, as we all know, China is a populace place. India, which also has its share of inhabitants, has increased emissions from a modest 1 ton to 1.7 tons per inhabitant. The USA, like Germany, has slipped down from 20.2 to 17.6 tons. From these figures you can certainly say that China holds the key. Then again, you can ask for whom China is producing – in any case, not solely for its own inhabitants. Today it clearly pays off to invest in China's rapidly growing renewable energy instead of Chinese coal.

The core argument for increasing growth is that curing the withdrawal from fossil energy is a costly premise, and that the only way we can obtain the means to support it is through economic growth. This argument builds on two myths at the heart of the question regarding how we approach the Anthropocene: through regression and a return to the past (the misanthropic and pessimistic or nostalgic alternative); or through allowing technology via the green shift to clean up the mess (the philanthropic, optimistic, and forward-looking alternative). Both ideas are mistaken, nor does the best alternative lie somewhere in the middle. Returning to the past is a utopian dream and not a desired one either. Throughout their development, humans have never looked back; on the contrary, a fundamental part of the human project is forward progress, change, innovation. These characteristics also gave us polymer chemistry and an endless number of ingenious carbon products. The future will come, but with a focus on quality rather than quantity. The quantity carousel has brought us all closer to the end of the world, and part of that myth is that our problems can be solved by more of the same actions that created these problems in the first place: more growth in order to solve the growth problem. That is something like taking out expensive consumer loans to cover the housing debt.

As it happens, Norway is not exactly destitute when it comes to being able to pay for any conversion costs. In fact, we have the world's largest state fund. At the moment of this writing, the Fund clocks in at more than $850 billion. More digits are not possible to give here because, even as I write these lines, the Fund has already changed by tens of millions. It moves up and down a bit, in keeping with oil prices and the stock market, but mostly it continues to climb. Some will argue that putting this money more actively into conversion is sound not only environmentally, but economically. This is especially so with a slightly longer perspective in mind, which, after all, represents the Fund's basic strategy, particularly if we realize that this $850 billion is the value of one moment's withdrawal from the carbon bank, an account that has built up over several hundred million years.

No matter the money's source, the argument still runs that reduced consumption is not only a naive and utopian ideal, but hostile to the climate, because it reduces our chances of freeing ourselves from petroholism without suffering hideous withdrawal symptoms. Yet the argument that we must borrow even more from the future in order to cover conversion costs builds on the idea that energy consumption, and all other consumption, will continue to rise. Consumption, the logic goes, will increase because the world's population will increase, perhaps even more than we imagined. The old truth that the world's population would flatten out somewhere short of 10 billion is giving way to new prognoses that point beyond 10 billion, indeed, all the way up to 13 billion. The world's poor, which comprises the majority of these 10 or 13 billion, must be aided. Once again, however, all prognoses are also based on the idea that consumption will increase. The idea has been put forward that Norwegian gas and oil offers the best means of rescuing the world's poor – and the world's climate – because it is greener than all other fossil energy. Such comfortable self-denial also prevents Norwegian fossil reserves from becoming stranded assets, simply because the country's oil and gas ought to be exempt from remaining untouched. These calculations, of course, do not add up, unless you believe in miracles. Those of us who are sceptical of miracles should ask what we personally can do to help the carbon cycle get back on the desired track – and even those who believe in miracles must do their part to help the miracle along.

Many aspects of modern life are astonishingly greedy in their generation of $CO_2$, and it is difficult to get a handle on the contributions our activities make, both large and small. As such, we should all be provided with access to a simple, free, personal $CO_2e$ calculator, together with straightforward user instructions. The good life can be reasonably carbon neutral. The logical course would be to cut our personal carbon emissions by a quarter, perhaps by promoting it as a competition that revolves around consuming less instead of more. It is true that 'people resist control', but here we are talking about motivating rather than controlling behaviour. It is not ill will that knocked the carbon cycle askew, but rather a set of circumstances set in motion by a 'system' that provides numerous incentives for consuming more, but very few for consuming less. As such, we all need a user manual outlining how we can reach the goal of cutting emissions, and that goal itself should have appeal. Still, the conservation idea often comes across as an eternal exercise in shouting 'no' and declaring what one is against. *No* to logging and industry, *no* to fun (*no* to fast cars and exotic holidays), *no* to oil and shopping, indeed, *no* to progress itself. Who or what has painted conservation into this corner? Is it the movement itself? If so, we need to find a better way of advocating for things to which we can say 'yes'. Yes, however, should not simply function as a hidden inversion of no, but rather the opposite. In this context Tesla has something to teach us: the sustainable alternative can be cool and progressive – something of which people want to be a part.

There is an aspect to this argument that finds its parallel in Justus von Liebig's uneven barrel staves (see Part 1). The point was that the shortest stave will always determine how much water a barrel can hold. For Robert Malthus, the shortest stave was food production, and he could not have foreseen how quickly that stave would grow in the wake of the green revolution. For that lack of foresight, he is now branded the 'world's first environmental pessimist'. Today we might say water forms the shortest stave and more atmospheric $CO_2$ is destined to shrink that stave even more in regions where water is already in short supply. Water shortage, of course, means food shortage.

In terms of economic growth, $CO_2$ neutral energy is the shortest barrel stave, although this stave naturally can and should be lengthened through a 'green' revolution, though in this case of a more technological kind. Should that happen, there will be a whole new generation of 'environmental pessimists' – and I could think of nothing better. It is certainly a chance I am willing to take. On the other hand, it is clear that Malthus was right *in principle*, as carrying capacity exists. Indeed, strictly speaking, there are multiple carrying capacities – one for each barrel stave. In this sense, we might say the carbon stave is 800 Gt, but there are also staves for water, food, land area and biological diversity, to name just a few, all of which are connected to the carbon cycle in various ways. The ultimate length of these staves and others is determined by the fact that our planet has a given capacity for biospheric production. If we could tap the energy from solar wind, for example, we would have an unlimited source of renewable energy minus the $CO_2$ costs as long as the sun continues to burn hydrogen. That would be fantastic, a stunning victory for human ingenuity. That would not mean unlimited growth, however, because we are still left with a single planet and those many other barrel staves. We cannot live on love and air alone (although we cannot live without them either), but neither can we purely consume energy.

Today the world's annual water usage is around 2,500 cubic km. That figure does not really say much by itself, though it has increased more than fivefold since 1900: 90 per cent of this usage goes toward food production, and predictions point toward a significant increase in demand, even as today's trends and IPCC forecasts regarding available freshwater resources indicate decline. Thus far the world's food production has kept pace impressively with the population increase, even if food availability suffers horribly from uneven distribution, but the latest IPCC forecasts also point to a decline in food production.

This has taken us away from the topic of carbon – or has it? We have spent the last century transforming the globe and the carbon budget in ways difficult to comprehend. Changes in the carbon cycle are not simply something we anticipate happening, but something we are already experiencing. It is not as if 'we have fifteen years left' or something similar: we have no time left at all. There is reason to be wildly optimistic about what technological innovation will bring, but there is reason to be wildly pessimistic if we believe technology will free us from the problem, or that someone (politicians) or something (nature) will fix the problem for us.

If any doubt remains, let us be clear: the carbon cycle is extraordinarily complex and it would be a miserable strategy to wait to take action until we fully understand the effects of all the switches and gears in this intricate and variable machinery. We have a management objective, the stabilization of the atmosphere's remaining carbon capacity – or $CO_2e$. Any doubt as to the need for this will do the planet no good. James Hansen, who was one of the first to calculate residual capacity in this context, believes that 2°C is beyond the comfort zone and even that is sufficient to marshal some of the feedbacks we have discussed. Indeed, even though we have not yet reached an average of two degrees warming, many people contend, with good reason, that they have begun to witness the effects of an angrier and more unsettled climate. The question, therefore, becomes how much each of us can emit, the maximum $CO_2e$ that each individual can allow themselves. Whereas today's reality largely dictates that we establish emission limits according to what we believe the economy permits, we must turn the tables and allow only so much economy (the old kind on a steady course) as the residual capacity permits.

The carbon cycle is a 'wild card', as David Archer put it, despite the fact that he is one of the people who knows the most about it. We have entered unknown territory with the historically unparalleled changes brought by the Anthropocene. Whatever uncertainty we face here should prompt anxiety and lead us to take action rather than becoming a source of comfort. Risk, as we know, is the product of probability and consequence, and although the probability for the worst case scenario is thankfully small, it is still enough to mean great risk, since the consequences are sky high. For what it is worth, it is estimated that the risk for ending up in the 1,000 ppm $CO_2$ range might be somewhere in the magnitude of 10 per cent. We take out total coverage on our house and car for the risk of a few in a thousand that we will suffer a fire or a total loss. Since $CO_2$ is invisible, and as hotter and longer summers still do not seem particularly threatening, we do not transfer that same rationale to the climate. The exceptions are those directly in the line of fire: those in the insurance industry itself or the owners of houses on flood-prone hillsides, for example, whose property must be replaced when 'century floods' occur every five years. No one knows, of course, how quick the snowball will travel, how large it will get and when it will stop. The snowball effect will also not lead to a snowball earth, but exactly the opposite, and the whole point here is to stop the ball in its tracks while it is still possible. It could be that the terrain and the snowball's nature will be such that it will not reach especially dangerous proportions, but we simply cannot take that chance. The principle of 'be prepared' has never been more critical.

Overall, though, we have reason to be proud of ourselves. It is the fruits of human genius that have brought insights into the atmosphere's composition and into carbon's many faces and uses. We have gained insight into the intricate paths of the carbon cycle, through cellular organelles, through chloroplasts and mitochondria, through food chains and ecosystems, in mountains, forests and oceans – and we have recognized along the way all the things we do not yet understand. For very good reason, we dislike plastic where it does not belong: in the ocean, on beaches and in the bellies of dying albatrosses. Taken of themselves, however, polymer chemistry and plastics are fabulous, but there is simply too much of them in the wrong places. In the same way, the ability to retrieve hydrocarbons from the earth's depths or from the ocean bottom represents fantastic engineering feats. Coal, oil and gas not only heat our homes, but are the basic pillars of modern society.

Our problem is that we are clever, but perhaps not clever enough. We recognize long-term consequences, but do not allow a long-term rationale to govern the short-term. We have evolved as problem-solvers and there is no reason our species will not still be here in one thousand, ten thousand or even one hundred thousand years. Meaning and time, furthermore, are inextricably tied. I find meaning in my own life, but only if I believe in the generations to come: the more generations, the more meaning. Therefore I also hope we will be here in a million years, even if evolution will have probably made some moderate adjustments to us. I will not finish by saying that all will certainly end well, however, because that is something we simply cannot know.

## AFTERLIVES

This story could also have been told from the perspective of the C atom, which whirled around the cosmos for billions of years before being sucked in by the gravity of young planet Earth four billion years ago. Here it made the rounds through blue-green bacteria, primitive cells, trilobites and psilophytes, took breaks in the deep ocean, and enjoyed constant trips through the atmosphere in companionship with two bonded Os, periodically also with four Hs or other, more infrequent partners, before entering the ecosystems through stomata and chloroplasts. We could have talked about labyrinthodontia, giant dragonflies and tyrannosaurs, mammoths and Neanderthals – and, finally, a *Homo sapiens*, myself for example.

If in the end – not of this book, but of life – I should totter up the hill overlooking my lake and reflect back upon the lake's and my own carbon budget, I will think that we have certain things in common. We both contribute to the atmosphere's $CO_2$ and $CO_4$ content. I contribute a bit more per year than the lake, but it has also been around for a few thousand years and will probably be there a few thousand more before being again overgrown and ending up a mire. Both the lake and I respirate carbon that was once bound by photosynthesis – and that is where the similarity ends. There are many thousand such lakes, but there are billions like myself. It is neither possible nor desirable to alter the lake's carbon cycle, but it is both possible and desirable to do something about the human carbon budget.

One can approach life's balance sheet in so many ways, with the moral balance perhaps seeming more critical than the carbon balance, though that too has a moral side. Despite my well-intended cycling and rare aeroplane trips, I have probably contributed around 9 tons of carbon annually. I have also contributed by having children, which is not a particularly carbon-neutral enterprise. A tiny fraction of what I emit is due to my inner thermostat, which has maintained my bodily temperature at 37°C and has supplied me with enough energy for an eventful life. More than 99 per cent of my carbon budget, however, will be tied to the products and services I have used, of which food is naturally a part. I have eaten steak and tomatoes in my life, and farmed salmon are not exactly neutral either. The carbon stored in my body, and which will one day be released as $CO_2$, is nothing in comparison to the carbon I have indirectly oxidized, even though I do all I can to minimize this quantity. Therefore I am not too worried about the 15 kg of carbon that I will eventually surrender to the atmosphere. In contrast, it would be a comforting thought to imagine that I will whirl on throughout history, a bit going into trees, a bit into the ocean, a bit into future generations, and so on, into eternity.

# REFERENCES

## PART 1: CARBON, CARBON EVERYWHERE

1 Linnaeus was perhaps the foremost exponent from a natural theological perspective, which holds that everything has a divine origin and therefore also a purpose. Nature, that is, was not simply a source for intense emotions, but carried a hidden pattern, a message, and a deeper meaning. Mankind's role, and Linnaeus' in particular, was to interpret that hidden message. Within that framework is also the idea that the Creator will do the cleaning up, for example when it comes to climate change. Of course, for good reason climate change was not something Linnaeus was worried about. More on this subject can be found in Dag O. Hessen, *Carl von Linné* (Oslo, 2000).

2 Charles David Keeling (1928–2005) was the first to register a notable increase of atmospheric $CO_2$ and the first who seriously warned against the possible consequences that this could entail. He has directly and indirectly inspired this and countless other books on climate and the carbon cycle. Fittingly, many biographies have been written about Keeling, but a short and discipline-oriented overview can be found in two articles: D. C. Harris, 'Charles David Keeling and the Story of Atmospheric $CO_2$ Measurements', *Analytical Chemistry*, 82 (2010), pp. 7865–70, and M. Heimann, 'Obituary: Charles David Keeling 1928–2005', *Nature*, 437 (2005), p. 331.

3 A good overview when it comes to this and more is Eric Roston, *The Carbon Age: How Life's Core Element Has Become Civilization's Greatest Threat* (New York, 2008). This is a wonderful exploration of carbon's time travel through the universe, its numerous forms and uses, leading to a discussion, as books

on carbon tend to do, on $CO_2$ and the climate. There are also many books that tackle the periodic system and include basic information on carbon, such as Hugh Aldersey-Williams, *Periodic Tales: The Curious Lives of the Elements* (London, 2011) (also published as *Periodic Tales: A Cultural History of the Elements, from Arsenic to Zinc* (New York, 2011)).

4 Carbon does not live alone, either in chemistry (in general) or biology, but occurs in company with other materials. The relationships between carbon and other key elements do not occur by chance, and the elements nitrogen and phosphorus in particular have a dominating position over carbon in many ways. The bible on this subject is the work of my good friends and colleagues Robert W. Sterner and James J. Elser, *Ecological Stoichiometry* (Princeton, NJ, 2002). For an updated overview article, see Dag O. Hessen, James J. Elser, Robert W. Sterner and Jotaro Urabe, 'Ecological Stoichiometry: An Elementary Approach Using Basic Principles', *Limnology and Oceanography*, LVIII/6 (2013), pp. 2219–36.

5 There are many books and countless articles on different aspects of the carbon cycle. One of the best, which unites readability with academic rigour, and also covers most of the subjects, methane included, is David Archer, *The Global Carbon Cycle* (Princeton, NJ, 2010). Many of Archer's discussions on the subject also exist as podcasts. A rather dated, but still good and valid overview of the carbon cycle's connections to other cycles can be found in P. Falkowski et al., 'The Global Carbon Cycle: A Test of Our Knowledge of Earth as a System', *Science*, 290 (2000), pp. 291–6.

6 Human evolution, both biologically and culturally, is probably more connected to fire than we have assumed. An updated history on human development, which also takes up this subject, is Yuval Noah Harari, *Sapiens: A Brief History of Humankind* (New York, 2015). This is also a book that largely addresses cultural evolution with carbon given a prominent place.

7 Richard Wrangham, *Catching Fire: How Cooking Made Us Human* (London, 2009).

8 Libby's own overview can be read in W. F. Libby, 'Radiocarbon Dating', *Chemistry in Britain*, v (1969), pp. 548–52.

9 Dating of cave art is an exciting – and contentious – subject.

Who were first, Asians or Europeans? Modern humans or Neanderthals? What does the art represent: metaphysical longing, rituals, prayers or simply the joy of painting? We will probably never know, and the subject is also an outlier in the story of carbon. A short but good overview of cave art dating can be found in H. Valladas et al., 'Paleolithic Paintings: Evolution of Prehistoric Cave Art', *Nature*, 413 (2001), p. 479. Otherwise, a broader discussion can be found in D. Whitley, *Cave Paintings and the Human Spirit: The Origin of Creativity and Belief* (Amherst, NY, 2009). As is often the case, Wikipedia also offers an accessible article, 'Cave Painting', http://en.wikipedia.org.

10 Jan Baptiste (or Jean-Baptiste or Johannes Baptist) van Helmont (1580–1644) was one of the early universal geniuses who paved the way for the Enlightenment (which is often dated from 1650 to 1800, even though there was naturally no absolute transition from the 'dark Middle Ages' to the Enlightenment era). He was a student of Paracelsus, one of the great alchemists who definitively had one foot in the Middle Ages and one foot in the Enlightenment. However, for our history, Helmont's description of $CO_2$ (which he called 'gas sylvestre'), together with his plant experiments, are the most relevant items, even though he also reached some false conclusions. See also E. Van den Bulck, *Johannes Baptist van Helmont* (Leuven, 1999).

11 Joseph Black (1728–1799) is among those described as the 'discoverer of $CO_2$'. How many people can actually have discovered this gas? Science is often characterized by significant discoveries that happen incrementally, and $CO_2$ can be 'discovered' and described with varying degrees of precision and insight. Sometimes parallel discoveries are made, and this was especially the case prior to lightning-quick publication channels or an Internet that broadcasts news to colleagues. Black, in any case, was undoubtedly among the discoverers of 'latent heat' (energy stored in, for example, coal or carbohydrates) and therefore a forerunner in the field of thermodynamics. For more on Black, and many other great scientific heroes up until our own day, see, for example, John Gribbin, *Science: A History, 1543–2001* (London, 2002).

12 It is possible that the term 'universal genius' might seem rather inflated throughout the text, but what else could one say about

a person like Joseph Priestley (1773–1804)? It is impossible to do him biographical justice here, nor is that the intent. Only one of Priestley's central works figures in our context: *Directions for Impregnating Water with Fixed Air* (London, 1772). A sample of his earlier works includes *The Rudiments of English Grammar* (1765), *A Chart of Biography* (1765), *Essay on a Course of Liberal Education for Civil and Active Life* (1765), *The History and Present State of Electricity* (1767) and *Essay on the First Principles of Government* (1768). It is true that there was less competition in academia, but if that is not universal genius, I do not know what is.

13 Antoine Lavoisier (1743–1794) is one of chemistry's central founders, transforming the discipline with the first actual chemistry textbook: *Traité élémentaire de chimie* (1789). A translated overview of some of his central texts related to carbon and oxygen can be found in Thomas Henry's translation of Lavoisier's *Essays, on the Effects Produced by Various Processes on Atmospheric Acid with a Particular View to an Investigation of the Constitution of Acids* (London, 1783). An accessible and more recent overview of Lavoisier and other great names within chemistry can be found in Matthew Daniel Eddy, William R. Newman and Seymour Mauskopf, *Chemical Knowledge in the Early Modern World* (Chicago, IL, 2014).

14 It must be acknowledged that Sweden has produced many great scientific names, many of them particularly significant, especially in the period between 1700 and 1900 when Sweden was truly a scientific powerhouse. Many of these figure in the history of carbon, notably Carl Wilhelm Scheele (1742–1786). Research into the archives of Scheele's widow unearthed Scheele's letter to Lavoisier and proved that Scheele's discoveries were earlier than either Priestley's and Lavoisier's. Scheele's work *Chemical Treatise on Air and Fire* (1777) showcases his earlier insights, but this was only a fraction of what he was working on, including articles on a significant number of the elements in the periodic table. He took a turn into organic chemistry when he wrote about different fats and lactose, as well as analysing the content of oxalic acid in rhubarb. Another universal genius? At least in chemistry's domain.

15 Jöns Jacob Berzelius (1779–1848) was one of Sweden's greatest minds and ranked alongside Lavoisier as one of the founders of

chemistry as a discipline (together with Robert Boyle and John Dalton). Berzelius helped introduce a number of new elements into the periodic table, and I believe he should be credited not only with describing these new chemical elements, but with describing the relationship between them (stoichiometry).

16 Friedrich Konrad Beilstein (1838–1906), born in St Petersburg but of German origin, was a disciple of Liebig and the inspiration for the *Beilstein Journal of Organic Chemistry*.

17 Carl Linnaeus' contribution to cataloguing was formidable, but it must be said that he came up short. No one today can say how many species exist and the concept itself is now problematic, especially when it comes to simple and single-celled organisms. To date about 1.7 million species have been described, but the total number that exists is perhaps 8 million or greater. More about Linnaeus' efforts can be found in Hessen, *Carl von Linné*, but for a more up-to-date account see, for example, Camilo Mora et al., 'How Many Species Are There on Earth and in the Ocean?', PLOS *Biology*, 23 August 2011, DOI: 10.1371/journal.pbio.1001127.

18 Organic chemistry's logic is accessible in any textbook on the subject. A book that offers both pleasure and frustration (pleasure when reading it, but frustration at having to cram all the cycles and reactions for exams) is Albert Lehninger's classic *Principles of Biochemistry* (1970), which couples organic chemistry with the biological carbon cycle's essential reactions. Many editions have been published since my student years, and the seventh edition (New York and Basingstoke, 2017) has significantly more to offer than my own second edition from the 1980s.

19 It is true that from our perspective diamonds last forever, at least if they are not subject to an extraordinary combination of pressure and heat. Eric Roston also visits that subject in *The Carbon Age*, but the Koh-i-Noor's history, as well as other myth-shrouded celebrities from the diamond world, are accessible in many forms, many originating with Edwin Streeter, *The Great Diamonds of the World: Their History and Their Romance* (London, 1882). In terms of synthetic diamonds and the modern uses of pure carbon, see Anke Krueger, *Carbon Materials and Nanotechnology* (Weinheim, 2010).

20 See Krueger, *Carbon Materials and Nanotechnology*.
21 Richard Buckminster Fuller (1895–1983) does not play a large role in our history, nor has he contributed to insights surrounding carbon chemistry, but his distinctive structures almost seem to suggest that he foresaw the discovery of the carbon form that would receive his name.
22 Harold W. Kroto et al., 'C60: Buckminsterfulleren', *Nature*, 318 (1985), pp. 162–3.
23 Biosphere 2 was probably the most extravagant, visionary and monumental experiment conducted on the face of the earth, a kind of ecological CERN. Whether it was realistic is a matter of debate and the experiment is mentioned in a variety of contexts. An inside description of the experiment is given in Jane Poynter, *The Human Experiment: Two Years and Twenty Minutes Inside Biosphere 2* (New York, 2008).
24 A good historical overview may be found in A. Ravve, *Principles of Polymer Chemistry* (New York, 2012).
25 A good discussion of August Kekulé can be found on Wikipedia: http://en.wikipedia.org.
26 For example, Lehninger, *Principles of Biochemistry*.
27 The German chemist Hermann Staudinger (1881–1965) certainly built on others' contributions, but he was also the first to recognize both the structure and usefulness of polymers. The essence of his landmark work 'Über Polymerisation', *Berichte der deutschen chemischen Gesellschaft*, LVI/6 (1920), pp. 1073–85, is a hypothesis that rubber and other polymers, such as starches, cellulose and proteins, consist of relatively small molecules that are linked together with covalent bonds to form long chains of macromolecules.
28 Much can be said about DuPont's many business ventures, but apropos the endless discussion of 'how the world will live after oil', it must be said that DuPont and many other companies have shown it is more fruitful to direct a solid portion of profits back into research than to fill shareholders' pockets. The fruits of DuPont's formidable polymer chemistry research intitiative are presented in a thorough and straightforward way at 'History of Du Pont's Nylon Fibers', CHA Inc., http://cha4mot.com, accessed 5 June 2017.
29 The history of the car comprises an integral part of carbon's

more recent history, mainly because cars are significantly responsible for converting fossil, organic carbon into $CO_2$, but also because the car has been the impetus for much of synthetic carbon chemistry (just look at the wheel). The history of the car, of course, has been told countless times and the personification of the auto industry through Henry Ford has also been amply covered, for example in Jonathan Glancey, *The Car: The History of the Automobile* (London, 2003).

30 Ian Locke, *The Wheel and How It Changed the World* (New York, 1995).

31 Norway actually has a few plastic pioneers; see Frode Weium, ed., *Volund 1999–2000: Plas i det moderne Norge* (Oslo, 2001); Kathrin Pabst, ed., *Plastic: Historier om plastbåten, Årbok Vest-Agder-museet* (Kristiansand, 2011); and Trine Nickelsen, 'Då plasten kom til Norge', *Apollon*, 9 April 2015, www.apollon.uio.no, accessed 5 June 2017.

32 Few things better symbolize the shift from technological optimism than plastic. From being a marvel and one of modernity's spearheads, it suddenly assumed, at least in the West, negative associations, labelling something as 'plastic' or garbage. A growing population's increasing consumption produces, among other things, enormous garbage piles, but whereas organic waste can be broken down and made useful as 'biogas' (methane), plastic is something that simply stacks up. For the subject of garbage in general see, for example, Thomas Hylland Eriksen, *Søppel: Avfall i en verden av bivirkninger* (Garbage: Waste in a World of Side Effects) (Oslo, 2011). The plastic problem, for its part, has been extensively documented in everything from YouTube snippets to blogs and academic articles, and is more academically covered in J. R. Jambek et al., 'Plastic Waste Inputs from Land and the Ocean', *Science*, 437 (2015), pp. 768–71.

33 Jambek et al., 'Plastic Waste Inputs from Land and the Ocean'.

34 The silent spring on bird mountain is discussed in Dag O. Hessen, 'Den tause våren', *Harvest*, 1 May 2011, http://harvest.as, accessed 5 June 2017.

35 Alan Weisman, *The World Without Us* (London, 2007).

36 Craig Venter's life and scientific merits cannot be described as other than legendary, or at least as singular and visionary; see www.jcvi.org.

37 See, for example, G. M. Church and E. Regis, *Regenesis: How Synthetic Biology Will Reinvent Nature and Ourselves* (New York, 2012).

38 Others besides James Lovelock (b. 1919) have naturally presented insights into the planet's gas composition, but Lovelock's contribution was foremost in connecting such insights with feedbacks and climate. His central work on the subject is James E. Lovelock, *Gaia: A New Look at Life on Earth* (Oxford, 1979). In an easier-to-read, illustrated version, Lovelock repeats his Gaia hypothesis and the central concepts upon which it is built. He also encounters a number of classic objections and largely ties all of his arguments to the global environmental condition: 'Mother Earth is Sick' and 'Gaia's State of Health', in *The Ages of Gaia: A Biography of Our Living Earth* (London, 1988). Much of the same, but from a more personal perspective, as well as an interesting discussion on natural scientific principles, God and Gaia, and holism versus reduction, can be found in James E. Lovelock, *The Revenge of Gaia* (London, 2006). This book also describes how we have surpassed the boundaries for global thermal regulation by our greenhouse gas emission and the terrifying consequences of this fact.

39 The person who truly gave us deeper insight into oxygen's development on earth was Robert Berner (1935–2015), who connects $CO_2$ and $O_2$ in his *The Phanerozoic Carbon Cycle: $CO_2$ and $O_2$* (Oxford, 2004). His student Donald Canfield is among those who have taken up the mantle, writing a number of revolutionary articles about the subject, sometimes together with Berner, and much of it is summed up in Donald E. Canfield, *Oxygen: A Four Billion Year History* (Princeton, NJ, 2014).

40 That ozone's development is closely connected to oxygen's is not surprising given that ozone is $O_3$. A simple overview of ozone and oxygen throughout the ages, but with a look at its correlation with life's evolution, is given in Dag O. Hessen, 'Solar Radiation and the Evolution of Life', in *Solar Radiation and Human Health*, ed. E. Bjertness (Oslo, 2008), pp. 123–36.

41 Much has been written on photosynthesis, and there is a nearly endless supply of scientific articles. For an overview, Canfield's *Oxygen: A Four Billion Year History* can also be recommended here. Most notably David Beerling provides an elegant overview

of photosynthesis and much more in his exceptional book *The Emerald Planet: How Plants Changed Earth's History* (Oxford, 2007). Beerling also demonstrates here that plants have been decisive for the planet's development on many levels, particularly through their absorption of $CO_2$, the production of oxygen and their ability to increase mountain erosion. Many of the primary articles related to $CO_2$, $O_2$ and erosion can be found in these two books. Otherwise, another good overview is Robert E. Blankenship, 'Early Evolution of Photosynthesis', *Plant Physiology*, 154 (2010), pp. 434–8.

42 Beerling also provides a good overview regarding the discussion of C3 and C4 plants and their evolution in relation to increasing $CO_2$ levels.

43 Thomas Robert Malthus (1766–1834) was the first to express concern openly regarding population growth in relation to resource base. He was qualitatively correct, but quantitatively in error. These issues have been pursued by the 'neo-Malthusians', though today population growth is the elephant in the room. Everyone knows it is a problem, but there is no direct way to tackle it. Part of the problem is that growth seems to be solved through an increase in affluence, at the same time as per capita consumption proves decisive for how far the load-bearing capacity can be exceeded (today we are far above it). More on this subject can be found in Thomas R. Malthus, *An Essay on the Principle of Population: Text, Sources, and Background Criticism*, ed. Philip Appleman (New York, 1976); D. H. Meadows, D. L. Meadows. J. Randers and W. W. Behrens III, *The Limits to Growth* (New York, 1972); D. H. Meadows, D. L. Meadows and J. Randers, *Beyond the Limits* (New York, 1991); and Jørgen Randers, *2052: A Global Forecast for the Next 40 Years* (White River Junction, VT, 2012).

44 Endosymbiosis was a decisive factor in life's development, as well as for the organelles that work to balance building and burning in the globe's carbon cycle. Lynn Margulis is the person credited with its discovery, although the Norwegian biologist Jostein Goksøyr conceived the idea at the same time. Of course, such endosymbioses are more common and more complex than most people realize, and have proven central for several branchings near the roots of life's tree: K. Shalchian-Tabrizi,

'Høiland Livets tre – og treets røtter', in *Mendels arv: Genetikkens æra* (Oslo, 2015), ed. Hessen, Lie and Stenseth, pp. 100–115.

45 Govindjee, J. T. Beattie, H. Gest and J. F. Allen, eds, *Discoveries in Photosynthesis* (Dordrecht, 2005).

46 Calvin's cycle is also called the Calvin-Benson cycle. Here again Lehninger's *Principles of Biochemistry* provides an overview. See also 'Calvin-syklus', UIO: Institutt for biovitenskap, 4 February 2011, www.mn.uio.no, accessed 5 June 2017.

47 For biological stoichiometry, see Sterner and Elser, *Ecological Stoichiometry*; and Hessen, Elser, Sterner and Urabe,'Ecological Stoichiometry: An Elementary Approach Using Basic Principles'.

48 Sterner and Elser, *Ecological Stoichiometry*.

49 Robert W. Sterner et al., 'Scale-dependent Carbon: Nitrogen: Phosphorus Seston Stoichiometry in Marine and Freshwaters', *Limnology and Oceanography*, LIII (2006), pp. 1169–80.

50 William G. Sunda and Susan A. Huntsman, 'Iron Uptake and Growth Limitation in Oceanic and Coastal Phytoplankton', *Marine Chemistry*, 50 (1995), pp. 189–206; A. J. Watson et al., 'Effect of Iron Supply on Southern Ocean $CO_2$ Uptake and Implications for Atmospheric $CO_2$', *Nature*, 407 (2000), pp. 730–33.

51 There is not lack of literature addressing these questions. In this context, a large number of books on the greenhouse effect will prove accessible. David Archer's *The Global Carbon Cycle* again provides a superb overview of the biogeochemistry in the carbon cycle, as well as an in-depth description of methane. The IPCC (Intergovernmental Panel on Climate Change) reports can be freely downloaded in various editions, though perhaps 'Summary for policymakers' can be considered an introduction. A personal book from perhaps the most central climate researcher in the last decades, James Hansen, is powerful reading: *Storms Over My Grandchildren: The Truth about the Coming Climate Catastrophe and Our Last Chance to Save Humanity* (London, 2009). If anyone should doubt that Hansen views the situation as critical, one need only glance at the subtitle.

52 For information on Tjeldbergodden Utvikling AS (TBU), see www.tbu.no.

53 William Martin et al., 'Hydrothermal Vents and the Origin of Life', *Nature Reviews Microbiology*, 6 (2008), pp. 805–14;

Kai-Uwe Hinrichs et al., 'Methane-consuming Archaebacteria in Marine Sediments', *Nature*, 398 (1999), pp. 802–5.

54 United States Environmental Protection Agency (EPA), 'Overview of Greenhouse Gases', originally available at www.epa.gov/climatechange/ghgemissions/gases/ch4.html. At the time of writing, this page has been pulled by President Donald Trump and Head of the EPA Scott Pruitt, now only available as a 19 January snapshot at https://19january2017snapshot.epa.gov/climatechange_.html.

55 IPCC 5th Assessment Report, www.ipcc.ch; Dana R. Caulton et al., 'Toward a Better Understanding and Quantification of Methane Emissions from Shale Gas Development', *Proceedings of the National Academy of Sciences*, CXI/7 (2014), pp. 6237–42.

## PART II: THE C IN CYCLE

1 Daniel C. Harris, 'Charles David Keeling and the Story of Atmospheric $CO_2$ Measurements', *Analytical Chemistry*, 82 (2010), pp. 7865–70; Charles D. Keeling and Minze Stuiver, 'Atmospheric Carbon Dioxide in the 19th Century', *Science*, 202 (1978), p. 1109.

2 Regarding Svante Arrhenius, see Spencer R. Weart, *The Discovery of Global Warming*, 2nd edn (Cambridge, MA, 2008). Two of Svante's many works are central: Svante Arrhenius, 'On the Influence of Carbonic Acid in the Air upon the Temperature of the Ground', *Philosophical Magazine and Journal of Science*, 5th ser., XLI/251 (1896), pp. 237–76; Svante Arrhenius, *Worlds in the Making: The Evolution of the Universe* (New York and London, 1908).

3 Regarding James Tyndall, see Weart, *The Discovery of Global Warming*.

4 V. S. Summerhayes and C. S. Elton, 'Contributions to the Ecology of Spitsbergen and Bear Island', *Journal of Ecology*, XI/2 (1923), pp. 214–86.

5 Raymond L. Lindeman, 'The Trophic-dynamic Aspect of Ecology', *Ecology*, XXIII/4 (1942), pp. 399–418.

6 The full story in strictly scientific terms can be found in Dag O. Hessen, Tom Andersen and Anne Lyche, 'Carbon Metabolism

in a Humic Lake: Pool Sizes and Cycling through Zooplankton.' *Limnology and Oceanography*, xxxv/1 (1990), pp. 84–9. On the same theme see Dag O. Hessen and Lars Tranvik, eds, *Aquatic Humic Substances: Ecology and Biogeochemistry* (Berlin and Heidelberg, 1999). See also Jonathan Cole et al.,'Plumbing the Global Carbon Cycle: Integrating Inland Waters into the Terrestrial Carbon Budget', *Ecosystems*, x/1 (2007), pp. 171–84.

7 E. Paasche, 'Coccholith Formation', *Nature*, 193 (1962), pp. 1094–5; U. Riebesell et al., 'Enhanced Biological Carbon Consumption in a High $CO_2$ Ocean', *Nature*, 450 (2007), pp. 545–8; Steven R. Emerson and John I. Hedges, *Chemical Oceanography and the Marine Carbon Cycle* (Cambridge, 2008).

8 Once again, for a good introduction to the carbon pump, see David Archer, *The Global Carbon Cycle* (Princeton, NJ, 2010), together with Steven R. Emerson and John I. Hedges, *Chemical Oceanography and the Marine Carbon Cycle* (Cambridge, 2008).

9 One (of many) studies on the importance of *Calanus finmarchicus* and its relative for fisheries and marine food webs can be found in K. T. Frank et al., 'Trophic Cascades in a Formerly Cod-dominated Ecosystem', *Science*, 308 (June 2005), pp. 1621–3.

10 More on the Gulf Stream and ocean climate can be found ibid.

11 The enormous subject of forests and carbon has led to heated debate among researchers, not to mention industry and conservation groups. What is the best approach and for what reason? Here we find many ecosystem services on a potential collision course: timber operations, other forest products, experiential value, biological diversity and, naturally, carbon sequestration. The following articles and discussions around the subject are relevant: Thomas H. DeLuca and Celine Boisvenue, 'Boreal Forest Soil Carbon: Distribution, Function and Modelling', *Forestry*, LXXXV/2 (2012), pp. 161–84; Erik Framstad et al., *Biodiversity, Carbon Storage and Dynamics of Old Northern Forests*, Nordic Council of Ministers (Copenhagen, 2013); Lars Gamfeldt et al.,'Higher Levels of Multiple Ecosystem Services are Found in Forests with More Tree Species', *Nature Communications*, 4 (2013), article no. 1340; Bjart Holtsmark, 'Harvesting in Boreal Forests and the Biofuel Carbon Debt', *Climatic Change*, CXII/2 (2012), pp. 415–28; Federico Magnani et al., 'The Human Footprint in the Carbon Cycle of Temperate and

Boreal Forests', *Nature*, 447 (207), pp. 848–52.
12 F. I. Woodward, 'Stomatal Numbers are Sensitive to Increases in $CO_2$ from Pre-industrial Levels', *Nature*, 327 (1987), pp. 617–18. See also Alistair M. Hetherington and F. Ian Woodward, 'The Role of Stomata in Sensing and Driving Environmental Change', *Nature*, 424 (2003), pp. 901–8.
13 Pauline Asingh and Niels Lynnerup, eds, *Grauballe Man: An Iron Age Bog Body Revisited* (Aarhus, 2007).
14 Statistics Norway, www.ssb.no/natur-og-miljo.statistikker/klimagassn/aar-endelige, 13 December 2016.
15 Regarding the function performed by these fungi, see for example Colin Averill, Benjamin L. Turner and Adrien C. Finzi, 'Mycorrhiza-mediated Competition between Plants and Decomposers Drives Soil Carbon Storage', *Nature*, 505 (2014), pp. 543–5.
16 For more on fungi in forests and their role in carbon sequestration, see for example K. E. Clemmesen et al., 'Roots and Associated Fungi Drive Long-term Carbon Sequestration in Boreal Forest', *Science*, 339 (March 2013), pp. 1615–18.
17 See www.regnskog.no; and 'Logging', *The Living Rainforest*, www.livingrainforest.org, accessed 6 June 2017.
18 Christopher E. Doughty et al., 'Drought Impact on Forest Carbon Dynamics and Fluxes in Amazonia', *Nature*, 519 (2015), pp. 78–82; Yadvinder Malhi, Richard Betts and Timmons Roberts, eds, 'Climate Change and the Fate of the Amazon', *Philosophical Transactions of the Royal Society B*, 363 (2008) (theme issue); Liming Zhou et al., 'Widespread Decline of Congo Rainforest Greenness in the Past Decade', *Nature*, 509 (2014), pp. 86–90; Thomas Hilker et al., 'Vegetation Dynamics and Rainfall Sensitivity of the Amazon', *Proceedings of the National Academy of Sciences*, CXI/45 (2014), pp. 16041–6.
19 Swamps and wetlands have traditionally been regarded as worthless before being drained and cultivated. It is only recently that we have realized what formidable carbon sinks swamps actually pose and the other ecosystem services they perform, for example, water storage. The example of Whittlesey Mere and other striking reminders of such ecosystem services can be found in Tony Juniper's prizewinning book *What has Nature Ever Done for Us? How Money Really Does Grow on Trees* (London,

2013). An overview of an evaluation of Norwegian ecosystem services can be found at 'Naturens goder – om verdier av økosystemtjenester', Norges offentlige utredninger 2013:10, www.regjeringen.no, accessed 6 June 2017.

20 Jødahlsmåsan is a striking Norwegian example where the carbon budget is estimated; see www.nrk.no/norge/_-blir-som-a-slippe-ut-co2-fra-150.000-biler-1.11990650.

21 James R. Arnold, Jacob Bigeleisen and Clyde A. Hutchison Jr, 'Harold Clayton Urey, 1893–1981', *Biographical Memoirs*, LXVIII (Washington, DC, 1995), pp. 363–411, available at www.nasonline.org, accessed 6 June 2017.

22 Valier Galy, Bernhard Peucker-Ehrenbrink and Timothy Eglinton, 'Global Carbon Export from the Terrestrial Biosphere Controlled by Erosion', *Nature*, 521 (2015), pp. 204–7.

23 See again Archer, *The Global Carbon Cycle*, and Roston, *The Carbon Age*.

24 Climate regulation as driven by vegetation, erosion, the connection of elements like carbon, nitrogen and phosphorus, and peculiar feedbacks is fascinating reading. See C. Buendia et al., 'On the Potential Vegetation Feedbacks that Enhance Phosphorus Availability: Insights from a Process-based Model Linking Geological and Ecological Timescales', *Biogeosciences*, 11 (2014), pp. 3661–83. James Lovelock's disciple Tim Lenton has taken the 'Gaia theory' quite a bit further in Timothy M. Lenton, 'Gaia and Natural Selection', *Nature*, 394 (1998), pp. 439–47, which gives an academic overview of the principles of the Gaia Theory and how it can be harmonized with standard biological conventions, including natural selection. From a purely scientific perspective, Lenton has developed Lovelock's biogeochemical feedback mechanisms: see Timothy M. Lenton, 'The Role of Land Plants, Phosphorus Weathering and Fire in the Rise and Regulation of Atmospheric Oxygen', *Global Change Biology*, 7 (2001), pp. 613–29; and Timothy M. Lenton and Andrew J. Watson, 'Redfield Revisited: 1. Regulation of Nitrate, Phosphate, and Oxygen in the Ocean', *Global Biogeochemical Cycles*, XIV/1 (2000), pp. 225–48. These feedbacks have also been increasingly included in the global climate models on which IPCC builds.

25 Yngve Vogt, 'Vil forvandle drivhusgass til stein', *Apollon*, 1 February 2015, www. apollon.uio.no, accessed 6 June 2017.

26 Paul Crutzen, 'Geology of Mankind', *Nature*, 415 (2002), p. 23.
27 Simon L. Lewis and Mark A. Maslin, 'Defining the Anthropocene', *Nature*, 519 (2015), pp. 171–80.
28 International Energy Agency, 'World Energy Outlook', 2014. Available at www.iea.org/newsroom/news/2016/november/world-energy-outlook-2016-html, accessed 9 August 2017.
29 For the World Coal Association, see www.worldcoal.org.
30 'Statistical Review of World Energy', available at www.bp.com/en/global/corporate/energy-economics/statistical-review-of-world-energy.html, accessed 9 August 2017.
31 A. R. Brandt et al., 'Methane Leaks from North American Natural Gas Systems', *Science*, 343 (2014), pp. 733–5.
32 For Lovelock's central works, see Part 1, note 38.
33 David Beerling, *The Emerald Planet: How Plants Changed Earth's History* (Oxford, 2007), is a masterly study.
34 Plenty of time series show that the forest is not only creeping northward, but that a systematic 'greening' is also happening in the Arctic regions. See, for example, Kevin C. Guay et al., 'Vegetation Productivity Patterns at High Northern Latitudes: A Multi-sensor Satellite Data Assessment', *Global Change Biology*, xx/10 (2014), pp. 3147–58.
35 Henrik Ibsen, *Brand: A Dramatic Poem in Five Acts* (1866), trans. John Northam, 2007, available at http://ibsen.nb.no, accessed 6 June 2017.
36 V. Ramanathan and G. Carmichael, 'Global and Regional Climate Changes Due to Black Carbon', *Nature Geoscience*, 1 (2008), pp. 221–3.
37 Ulf Riebesell, 'Climate Change: Acid Test for Marine Biodiversity', *Nature*, 454 (2008), pp. 46–7; O. Hoegh-Guldberg et al., 'Coral Reefs under Rapid Climate Change and Ocean Acidification', *Science*, 318 (2007), pp. 1737–42; James C. Orr et al., 'Anthropogenic Ocean Acidification over the Twenty-first Century and its Impact on Calcifying Organisms', *Nature*, 437 (2005), pp. 681–6.
38 Aragonite and calcite also have different saturation points relating to $CO_2$ and pH, and those creatures equipped with an aragonite shell are worse off. Most articles on acidification make this point, but a good overview can be found here: Scott C. Doney et al., 'Ocean Acidification; The Other $CO_2$ Problem',

*Annual Review of Marine Science*, 1 (2009), pp. 169–92.

39 Chris Langdon et al., 'Effect of Calcium Carbonate Saturation State on the Calcification Rate of an Experimental Coral Reef', *Global Biochemical Cycles*, XIV/2 (2000), pp. 639–54; M. J. Atkinson et al., 'The Biosphere 2 Coral Reef Biome', *Ecological Engineering*, XIII/1–4 (1999), pp. 147–71.

40 Regarding the warm period 55 million years ago, see again Archer (*The Global Carbon Cycle*), Beerling (*The Emerald Planet*) and Canfield (*Oxygen: A Four Billion Year History*).

41 In Elizabeth Kolbert's Pulitzer Prize-winning book *The Sixth Extinction* (New York, 2014), the author describes a fascinating diving tour she took to visit the vents. Ocean acidification, as we see, poses a significant threat to ocean life. Regarding climate effects in terms of the ocean in general, see Geoffrey K. Vallis, *Climate and the Oceans* (Princeton, NJ, 2012).

42 Joel E. Johnson et al., 'Abiotic Methane from Ultraslow-spreading Ridges Can Charge Arctic Gas Hydrates', *Geology*, XLIII/5 (2015), pp. 371–4.

43 Archer, *The Global Carbon Cycle.*

44 Paul F. Hoffman et al.,'A Neoproterozoic Snowball Earth', *Science*, 281 (1998), pp. 1342–6; Robert E. Kopp et al., 'The Paleoproterozoic Snowball Earth: A Climate Disaster Triggered by the Evolution of Oxygenic Photosynthesis', *Proceedings of the National Academy of Sciences*, CII/32 (2005), pp. 11131–6.

45 Petit, J.R. et al., 'Climate and Atmospheric History of the Past 420,000 Years from the Vostock Ice Core, Antarctica', *Nature*, 399 (1999), pp. 429–36.

46 Steve Graham, 'Milutin Milankovitch (1879–1958)', NASA *Earth Observatory*, http://earthobservatory.nasa.gov, accessed 6 June 2017.

47 Mark Pagani et al., 'An Ancient Carbon Mystery', *Nature*, 314 (2006), pp. 1556–7. Henrik Svensen et al., 'Release of Methane from a Volcanic Basin as a Mechanism for Initial Eocene Global Warming', *Nature*, 429 (2004), pp. 542–5. PETM is also thoroughly discussed in Archer's *The Global Carbon Cycle*.

48 Svensen et al., 'Release of Methane from a Volcanic Basin as a Mechanism for Initial Eocene Global Warming', p. 542.

49 Henrik Svensen, 'Geochemistry: Bubbles from the Deep', *Nature*, 483 (2012), pp. 413–15.

50 Robert Berner, *The Phanerozoic Carbon Cycle: $CO_2$ and $O_2$* (Oxford, 2004); Canfield, *Oxygen: A Four Billion Year History*; Colin Goldblatt, Timothy M. Lenton and Andrew J. Watson, 'Bistability of Atmospheric Oxygen and the Great Oxidation', *Nature*, 443 (2006), pp. 683–7.
51 Stephen Jay Gould, *Wonderful Life: The Burgess Shale and the Nature of History* (London, 1990).
52 Ian J. Glasspool and Andrew C. Scott, 'Phanerozoic Concentrations of Atmospheric Oxygen Reconstructed from Sedimentary Charcoal', *Nature Geoscience*, 3 (2010), pp. 627–30.
53 The story of Charles Brongniart's discoveries is well described in Beerling's *The Emerald Planet.*
54 J. T. Randerson et al., 'Global Burned Area and Biomass Burning Emissions from Small Fires', *Journal of Geophysical Research: Biogeosciences*, 117, DOI: 10.1029/2012JG002128 (2012).
55 Nicolas Gruber and James N. Galloway, 'An Earth-system Perspective of the Global Nitrogen Cycle', *Nature*, 451 (2008), pp. 293–6.
56 Karl A. Wyant, Jessica R. Corman and James J. Elser, *Phosphorus, Food, and Our Future* (Oxford, 2013). The authors believe phosphorus is in danger of becoming a critically limited resource and there is much to indicate they are correct. The green revolution that proved Malthus wrong can be reversed if we experience a critical phosphorus shortage in the future. See also 'Sustainable Phosphorus Alliance', https://phosphorusalliance.org, accessed 6 June 2017.
57 James J. Elser and Dag O. Hessen, 'Biosimplicity via Stochiometry: The Evolution of Food Web Structure and Processes', in *Aquatic Food Webs: An Ecosystem Approach*, ed. Andrea Belgrano et al. (Oxford, 2005).

## PART III: THE FOOTPRINT

1 Yuval Noah Harari, *Sapiens: A Brief History of Humankind* (New York, 2015); Torfinn Ørmen, *Historien om oss* (Oslo, 2012).
2 See World Wild Life Fund, 'Living Planet Report 2014', www.worldwildlife.org, accessed 6 June 2017.
3 Elizabeth Kolbert, *The Sixth Extinction* (New York, 2014).

4 For Global Footprint Network and Earth Overshoot Day, see www.footprintnetwork.org.
5 IPCC 5th Assessment Report, www.ipcc.ch.
6 Tim Flannery, *Here on Earth* (Melbourne, 2010); James Hansen, *Storms of My Grandchildren: The Truth about the Coming Climate Catastrophe and Our Last Chance to Save Humanity* (London, 2009); Spencer R. Weart, *The Discovery of Global Warming*, 2nd edn (Cambridge, MA, 2008).
7 Arne Johan Vetlesen, 'Ødeleggelsestvangen', *Klassekampen*, 17 January 2015, www.klassekampen.no. accessed 6 June 2017.
8 Mike Berners-Lee, *How Bad are Bananas? The Carbon Footprint of Everything* (Vancouver, 2011). On consuming, see also Oksana Mont et al., *Förbättra nordiskt beslutsfattande genom att skingra myter om hållbar konsumtion*, TemaNord 2013:552 (Copenhagen, 2013).

# BIBLIOGRAPHY

Aldersey-Williams, Hugh, *Periodic Tales: The Curious Lives of the Elements* (London, 2011)

Alfsen, Knut, Dag O. Hessen and Eystein Jansen, *Klimaendringer i Norge: Forskernes forklaringer* (Oslo, 2013)

Archer, David, *The Global Carbon Cycle* (Princeton, NJ, 2010)

Beerling, David, *The Emerald Planet: How Plants Changed Earth's History* (Oxford, 2007)

Berners-Lee, Mike, *How Bad are Bananas? The Carbon Footprint of Everything* (Vancouver, 2011)

Canfield, Donald E., *Oxygen: A Four Billion Year History* (Princeton, NJ, 2014)

Emerson, S. R., and J. I. Hedges, *Chemical Oceanography and the Marine Carbon Cycle* (Cambridge, 2008)

Flannery, Tim, *Here on Earth* (Melbourne, 2010)

Hansen, James, *Storms of My Grandchildren: The Truth about the Coming Climate Catastrophe and Our Last Chance to Save Humanity* (London, 2009)

Harari, Yuval Noah, *Sapiens: A Brief History of Humankind* (New York, 2015)

Juniper, Tony, *What Has Nature Ever Done for Us? How Money Really Does Grow on Trees* (London, 2013)

Kolbert, Elizabeth, *The Sixth Extinction* (New York, 2014)

Krueger, Anke, *Carbon Materials and Nanotechnology* (Weinheim, 2010)

Lovelock, James E., *Gaia: A New Look at Life on Earth* (Oxford, 1979)

Roston, Eric, *The Carbon Age: How Life's Core Element has Become Civilization's Greatest Threat* (New York, 2008)

Vallis, Geoffrey K., *Climate and the Oceans* (Princeton, NJ, 2012)

Weart, Spencer R., *The Discovery of Global Warming* (Cambridge, MA, 2008)

Wrangham, Richard, *Catching Fire: How Cooking Made us Human* (New York, 2009)

# ACKNOWLEDGEMENTS

This book has benefited from various research projects, and I am very much in debt to my colleagues on these projects, and in particular to my good friend and colleague Tom Andersen who has been instrumental in the development of ecological stoichiometry, carbon cycling and beyond. I am also indebted to my Norwegian publisher, Cappelen Damm, and my editor Erik Møller Solheim, and to Reaktion Books and Editorial Director Vivian Constantinopoulos for their devoted efforts at all stages of writing and translation.

Above all, thanks to my children for motivating my concern for the future and to my father, my great inspiration, the best I could have had, who passed away as this book was completed.

Carbon is a fascinating element, but as we all – at least almost all of us – realize these days, it is also potentially dangerous. To understand more of its complex cycle, and how we humans affect it, is literally a burning issue.

# INDEX

Page numbers in *italic* refer to illustrations